THE FUTURE ROLE OF BIOFUELS IN THE NEW ENERGY TRANSITION

DISCLAIMER

Any opinions, findings, and conclusions or recommendations expressed in this material are those of the author(s) and do not necessarily reflect the views of the participating parties. The results, analysis, conclusions or recommendations expressed in this book are based upon research projects, published documents, technical presentations, and data gathering conducted by the authors. Finally, the opinions, hypothesis, and conclusions or recommendations expressed in this book are the responsibility of the authors and do not necessarily reflect the vision of FAPESP.

ACKNOWLEDGEMENTS

The authors of this book would like to express their gratitude to the São Paulo Research Foundation (FAPESP) and Editora Blucher for financially supporting this publication. Also, we would like to thank FAPESP for supporting the research projects associated with this publication:

- Diretrizes de Políticas Públicas para a Agroindústria Canavieira do Estado de São Paulo (Process Number 2006/51725-1);
- Bioenergy Contribution of Latin America & Caribbean and Africa to the GSB project – LACAf-Cane Thematic Project (Process Number 2012/00282-3);
- Universidades e empresas: 40 anos de ciência e tecnologia para o etanol brasileiro (in Portuguese) (Process Number 2015/22108-3);
- Technological Roadmap for Sustainable Biofuels for Aviation: the opportunities for Brazil (Process Number 2012/50009-1).

The Authors

FAPESP

Blucher

Luís Augusto Barbosa Cortez
Frank Rosillo-Calle

THE FUTURE ROLE OF BIOFUELS IN THE NEW ENERGY TRANSITION

Lessons and perspectives of biofuels in Brazil

The future role of biofuels in the new energy transition: lessons and perspectives of biofuels in Brazil

Editora Edgard Blücher Ltda.

Publisher Edgard Blücher
Editors Eduardo Blücher and Jonatas Eliakim
Production manager Andressa Lira
Graphic designer Guilherme Henrique
Cover images Fermentec and Dedini
Cover designer Leandro Cunha

Blucher

Rua Pedroso Alvarenga, 1245, 4º andar
04531-934 – São Paulo – SP – Brasil
Tel.: 55 11 3078-5366
contato@blucher.com.br
www.blucher.com.br

Cataloging in Publication (CIP)
Angélica Ilacqua CRB-8/7057

Cortez, Luís Augusto Barbosa

The future role of biofuels in the new energy transition : lessons and perspectives of biofuels in Brazil / Luís Augusto Barbosa Cortez, Frank Rosillo-Calle – São Paulo : Blucher, 2023.

184 p.

Biography

ISBN 978-65-5506-746-0

1. Biomass energy 2. Biotechnology 3. biofuels I. Title II. Rosillo-Calle, Frank

23-1660 CDD 636.9

Table of contents

Preface

Air, water, and energy are essential ingredients of human life. The energy needed to sustain human life is quite modest: 2000 kilocalories per day (0.1 kilogram of oil equivalent) which was the daily energy consumed from food by human beings in the Neolithic era. Today it is 5 times higher in the least developed and 50 times higher in the high-income countries. With a world population of more than 7 billion people enormous amount of energy are required: approximately 13 billion tons equivalent of oil per year.

Up to the 18th century most of the energy used was wood from native forests and crop residues as the main source of heat for cooking and heating ("traditional renewables").

In the 19th century with the invention of the Watt engine and its use for electricity generation the importance of coal increased substantially replacing traditional fuels. Electricity generated in hydropower plants contributed also since 1880. The use of traditional energy sources became restricted to rural areas and less developed countries.

By the end of the 19th century, the use of oil increased and the advent of the automobile made oil the dominant fuel of the 20th century. Natural gas usually associated with oil production contributed to the gradual phasing of coal.

The growth of energy consumption in the 20th century was spectacular and much faster than population growth: from 1850 to 2020, the world population increased 6 times (1.238 billion people to 7.760 billion). Energy consumption grew 28 times (from 486 million to 13.527 billion): 83.7 from fossil fuels and 16.3% from "new renewables" (biomass, wind, solar, geothermal) hydro and nuclear.

The drivers of these changes were technological developments, costs, and access to energy particularly oil which was determined in a large scale by the policies of OPEC (Organization of Petroleum Exporting Countries) formed by the main oil producers in the world the most important being Saudi Arabia, United Arabia Emirates, Kuwait, Iraq, Iran, Nigeria, and Venezuela. Environmental considerations were also very important

In Brazil, petroleum derivatives always represented a large share of energy consumption – approximately 40% – due to the fact that since the middle of the 20th century the Government privileged road transportation.

Today Brazil is a highly motorized country (392 cars/1,000 people compared to 816 cars/1,000 people in the US) but its railway system 5 times smaller than the US system (0.147 km/1,000 people compared to 0.752 in the US.

In 1975, local oil production was very small, and gasoline and diesel oil imports posed an enormous burden on the economy representing roughly one-half of the earnings from exports. At the same time, the sugarcane industry was facing a serious crisis due to declining prices of sugar in the international market. The Government decided to stimulate through subsidies, the production of large amounts of ethanol from sugarcane, which was known to be a good replacement for gasoline. It was thus launched the ETHANOL PROGRAM.

Today approximately 30 billion litres of ethanol per year from sugarcane are produced in Brazil replacing, at competitive prices, approximately one-half of the gasoline that would otherwise be in use. In addition to that, bagasse from sugarcane contributes with 9% of the electricity production of the country.

The ETHANOL PROGRAM led to the development of the important industrial of the production of refineries. The productivity of ethanol production threefold in 50 years and "cradle to grave" analysis showed that ethanol from sugarcane produced carbon emissions substantially smaller than gasoline.

As we enter the 21st century further efforts in replacing fossil fuels at the world level are becoming an urgent task due to the increased concerns on the consequences of global warming.

There are many options for that transition such as the use of "green hydrogen" long life batteries for electric vehicles, carbon capture and storage and others. Government policies can favor any of them through subsidies and/or reallocation of government funds. The successful Brazilian ETHANOL PROGRAM is an excellent example of how to it.

This book by Luís Cortez and Frank Rosillo-Calle describes the origins of the PROGRAM its development, the several challenges faced and how they were successively met. It will be very helpful to people working on the new frontier of a sustainable energy future.

Professor José Goldemberg
University of São Paulo (USP)

Avant Propôs

The motivation to publish this book has been to discuss the main challenges facing the global energy transition and how the Brazilian biofuels are affected.

Many people, including energy experts and the media, believe biofuels still represent an old model, with many negative effects such as competing with food production and threatening ecological sanctuaries like the Amazon Forest. Unlike biofuels, electric vehicles are regarded as modern, efficient, and a cleaner alternative. This view is misguided and denote ignorance. With such attitudes, the obvious alternative leads to a substitution of biofuels for electricity. However, things are not black and white and here, in this book, we try to unveil a more realistic scenario.

First, Brazil, which is still a young democracy, struggled and succeeded to create a fossil fuel alternative that was the envy of the world for its positive impacts. For example, improvement of energy independence, creation of large number of jobs (both rural and industrial) and more importantly, the ethanol program demonstrated to be one of the few successful endogenous initiatives put together by government, entrepreneurs, and society. This great "exercise" helped Brazil to have the cleanest energy matrix among the ten largest world economies.

Second, the introduction of electric vehicles in Brazil could result in a negative balance, as far as the GHG emissions are concern. Why? Because ethanol produced in Brazil has already very low carbon footprint and there is no guaranteed that electricity, necessary to power the new vehicle fleet, will result in any environmental benefit. Large scale use of electricity poses serious problems, as stated in the book.

Brazil, the largest tropical country in the world, has much better ways to contribute to the global effort to reduce GHG emissions. Among the options, for example and as discussed in the book, is land use reform in which pastureland could be improved, resulting in considerable positive economic and environment benefits. For example, reforestation and regrowth of part of the Amazon Forest ecosystem, will lead to more land be available for sustainable food production for a hungry world.

In this book, the authors try to demonstrate that Brazil should maintain and expand its biofuels program by ensuring more sustainable land use, and by investing in new innovating engine technologies, such hybrid vehicles and fuel cells, besides more modern process technology and sustainable feedstocks

Objectives

The energy sector is in turmoil with multiple possible scenarios, with huge economic, social, political, environmental, and technological implications. The big question is, what could be the potential role of biofuels for transport in this emerging and mixed energy scenario, and more specifically in Brazil? In the very short term, this uncertainty will translate in greater increase of oil and gas and renewables (RE) in general, as there are few realistic alternatives. But in the longer term, it must be non-fossil fuels. The emerging global consensus is that the future MUST be renewables. The question is, how long will take this transition phase?

The Russian war on Ukraine represents a major energy and geopolitical shift, although such impacts will be unequal around the world. For example, and as stated in chapter 1, in Europe given its high dependency on Russian fossil fuels there is a political scramble to reduce such dependency as a matter of urgency, which could be translated in an urgent search for energy alternatives, especially RE. Other countries are also increasing RE but as a response to high oil prices e.g., India which is increasing gasoline-ethanol blend to reduce costly oil imports. Thus, there will be a variety of technological, economic, and energy alternatives and policy decisions, depending on the country.

During the oil crisis of the 1970s, Brazil showed a great vision by setting up a national ethanol fuel program, a unique project at national level, that served as a school and was envy of many countries.

This initiative put Brazil in a unique historical footage e.g., a considerable know-how on alternative fuel for transport, as well technological, agricultural, economic, environmental, and social benefits. Thus, the key question we need to ask is what could be the new role of biofuels in the emerging energy paradigm in the country? What lessons can be applied from its unique historical experience? What lessons are there for other countries?

The book examines such questions in detail and tries to provide answers as accurate as possible within this highly uncertain future scenario. Science and technology are advancing so quickly, that options that seem unfeasible today could become a reality tomorrow. And even technologies seem feasible today, could easily be obsolete tomorrow.

The main challenge for Brazil is political rather than technological, social, or economic, since the country has the scientific and technological know-how, natural and human resources, and research capability to deal successfully with various energy future scenarios. But there are clearly some important scientific and technological gaps, and the need to rethink present economic biofuels-related models that will require political actions. When this is the case, the corresponding recommendations are made.

This book will try to answer some of these questions particularly with a wide readership in mind. To inform the international community about the historical development of biofuels in the country, the new emerging reality, and perspectives and potential impacts of new biofuels scenarios.

Introduction

The energy sector is going though considerable upheavals for a combination of factors e.g., climate change, environmental pressure, decarbonization, and the Russian war on Ukraine. The time where little used to happen is gone. Take, for example the oil sector, and more specifically transport, which has not been synonymous of rapid change in the past. In fact, for decades, hardly any significant change took place. For decades, the automobile industry and oil and gas giants, have shared a married of convenience.

The undergoing changes are unparalleled. The energy sector is generally very volatile, and this situation is further compounded by the Russia-Ukraine war. Oil volatility is rife, with market dislocation increasing. This, together with surging energy prices will curb demand, at least in the short term.

There are many "unknowns" but undoubtedly such changes represent a paradigm shift with cataclysmic proportions. Huge economic, social, environmental and policy implications, with hundreds of billions of dollars are at stage. Such changes represent both huge opportunities and also unprecedented challenges and pitfalls.

What is then the likely energy scenario in the near future? This is clouded with uncertainty. There are not clear enough energy trends to predict, with reasonably degree of certainty, what may happen in the next 10-20 years, not to mention beyond this. There have been many scenario predictions but hardly any have turned out to be correct. Let focus on transport. There is no question that the automobile of the future will be very different, in which decarbonization, environment and sustainable fuels will play a key role.

The energy transition will be largely conditioned to advances is decarbonization, which, at the same time, will be affected by many and diverse factors, e.g., political, economic, environmental, social, and technological. Such changes will be uneven throughout the world. This involves a multitude of mutually supporting measures. Climate change abatement requires a minimum of global cooperation. Unfortunately, this is an area characterised by huge differences between the most advanced economies versus the poorest ones. The replacement of fossil fuels by renewable energy (RE) will be uneven around the world. For example, in some countries electric vehicles are already well underway, as are biofuels.

These impinging changes represent an enormous challenge to both the automobile and oil and gas industries, as stated. While oil and natural gas are not going away any time soon it will, progressively, be phased out over time. And many candidates are progressively emerging e.g., electricity, hydrogen, solar, biofuels (bioethanol, biodiesel). In the future there will be a mixture of fuels in which none will be playing the dominant role of oil. The question is which ones will prevail in this arras of possible alternatives. Only time will tell.

At the time of writing, it is impossible to predict with any degree of certainty what will be the real impact of Russian war on Ukraine, except that the impact will be felt for a long time and unequally around the world. The old system is being replaced without a clear vision of what will be put in its place. This is particularly the case in Europe with is highly dependent on Russia for fossil fuels and is urgently trying new sources and alternatives. Time will tell.

But at the same time, there are reasons for optimism. Human ingenuity has almost no limits. We are facing huge problems, but never in history are so many researchers and resources dedicated to find solutions. New technologies and innovations are advancing exponentially. A clear example has been the search for the COVID-19 vaccines, which in normal times would have taken years to develop, but in reality, took months. Another example is UN COP26 in Glasgow and COP27 in Egypt. And a further example of international cooperation is the UN Convention on Biodiversity, 7-19 December 2022, in Montreal, that has agreed to Adopt a Global Biodiversity Framework. Despite huge difficulties, it has been possible to reach some fundamental global agreements, that affect the global community. Of course, it remains to be seen how this will translate in fundamental and real actions.

It shows that when Homo Sapiens confront a serious problem are capable of finding solutions. Without forgetting the human stupidity and capacity for self-destruction. With the energy sector could be the same. After all, the most complex thing we know in our universe is the "human brain". A simple, immediate, and cheap solution is for all of us (particularly those who consume a lot and waste even more) to use less energy, without negatively affecting our living standards. We all can make choices and many small choices amount to a big one. As Gandhi, said: *There is enough for everybody needs but not everyone's greed.* The new reality requires bold decision-making and political brinkmanship!

This book is concerned with the potential role of biofuels in transport, one of the alternatives being pursued within this changing energy scenario, with Brazil as the key focus. Why biofuels? Because they have been around for many decades, and in particular in Brazil and USA, and to a lesser extent in many other countries. They represent a proven alternative, be it limited. Biofuels represent little technological challenge by comparison, as it is a well proven technology and fuel. The pros and cons of biofuels have been heatedly debated, but on balance and, if produced and used sustainably, the benefits far aweigh the negative impacts. Therefore, biofuels represent a considerable advantage over other emerging alternatives.

Biofuels have been around for decades and are an integral part of the fuel mix in the transportation systems. Currently this sector provides between 3-4% of fuel consumption in road transportation, with USA and Brazil as the main producers and consumers, and to a lesser scale in the EU, China, and India, among others.

Unlike petrol, the use of biofuels is patchy, and their contribution on a global scale is unequal, representing a minuscule contribution, or none al tall, in many countries. This uneven distribution reflects resource endowment, and policy options, cultural factors, lack capital, know-how, etc. The totality of biofuels is used blended with petrol and diesel in varying proportions. But we cannot forget that biofuels, particularly ethanol, have multiple other industrial uses.

The question we are posing here is, the extent to which biofuels are likely to play in a rapidly changing energy scenario? To what extend the paradigm shift will enhance or inhibit their development, as oil is gradually replaced by other emerging alternatives, in addition to biofuels, e.g., electricity, hydrogen, solar, etc? And most importantly, what role can biofuels play in Brazil in the future?

Growing concern with climate change, environment, and sustainability, together with social and political pressures, means that this paradigm shift cannot be reversed. On the contrary, it will accelerate. Climate change is a growing concern and the huge potential impacts have not yet fully penetrated into the majority of the population psychic, who is struggling with everyday life and regard climate change as something that may happen in the future.

But biofuels have also negative effects that limit their potential. The main one being the perception of land competition with food production and the requirement of being environmentally sustainable. The "food versus fuels" dilemma has been the object of heated debate for decades. The debate has been clouded by vested interest, bias, and lack of reliable scientific data. The "food versus fuel" debate needs to consider the intertwined nature of biofuels, land use, food production, food waste, animal feed production, diets, social habits, food distribution systems, policy, etc. In fact, the FvF argument is being discredited, even in the EU where there has been a strong debate (see Chapter 6).

The biofuel industry needs to transform itself and get rid of the FvF argument, even in Brazil where this has not been a serious issue. Biofuels production needs to be complementary to food production.

Biofuels have still mountains to claim, but new technological developments e.g., converting cellulose-based biomass to ethanol, will make available large additional amounts of feedstocks in many more countries, at competitive cost in the near future and on a sustainable manner.

What then are the potential opportunities and challenges confronting biofuels in this rapidly changing energy scenario? Should biofuels be promoted as a vehicle fuel or should the focus be on new emerging markets e.g., aviation, maritime transport, or fine chemicals? This will depend on the specific circumstances. The book investigates such potential alternatives with focus in Brazil.

Biofuels will not be the answer to oil, but they can play a much greater role as it has been the case so far in the fuel mix. This is particularly so in the Brazilian case. Understanding such trends will provide a better world view on these changes. And just keep in mind that there is not, quite simply, a 100% pollution-free fuel; just fuels than pollute more than others.

Five central themes are the main focus of this book.

Firstly, it provides an overall global view of the energy sector and the changing scenarios and the role of biofuels within it. No country is an island and what happen in the world will have an impact in Brazil, directly or indirectly, and in a minor or major scale, and vice versa. This overview will help you to better understand the current and potential directions of the energy sector, primarily transport.

Secondly, the book examines the history of biofuels development in Brazil. The history of ethanol fuel is fascinating, with many twists and turns. No large program is exempted of human desires, views, vested interests, political infightings, and so forth. There are important lessons from this unique historical experience.

The third theme describes in great detail the present situation of biofuels in Brazil (ethanol and biodiesel), ranging from current status, opportunities and challenges facing this industry and wider implications. Major changes are underway e.g., the rapid expansion of corn for ethanol production, challenging the dominant role of sugarcane, or the utilization of soybean oil for biodiesel production. The potential changes in the transportation system with new fuels e.g., electric cars, hydrogen, etc., are leading to new scenarios.

Theme four focuses on the fundamental R&DD required to overcome the challenges. It demonstrates that Brazil has considerable R&DD capacity and know-how to develop new feedstocks, particularly from woody biomass, and process technologies.

The final theme (Chapter 5) focuses on the future of biofuels in Brazil and try to answer the fundamental questions posed in the book. It seems that a key issue will be political rather than anything else.

And finally in chapter 6, we summarize the key points covered in the book. This chapter provides a succinct and quick overview, together with final thoughts.

Why this book, the reader might ask? This book has been born from the desire to explain to the non-expert in biofuels, and in particular the international community, of their importance. To share the authors rich experience over many years with the development of biofuels, and to provide a better understanding. It is aimed at a wide readership, from the policymaker to the general public.

We live in a world inundated by scientific data, rapid technological change, misinformation, (often beyond human comprehension), that confuse the reader; a world where the only thing we all seem to agree and matters more (between friends and enemies) *is money*. Throwing a bit of light should help the reader to better understand the world of biofuels and their merits and implications.

CHAPTER 1
BRIEF GLOBAL ENERGY OVERVIEW: TRENDS/SCENARIOS AND THE NEW ROLE OF BIOFUELS

1.1. INTRODUCTION

The aim of this chapter is to provide a brief global overview of energy developments (trends and scenarios) in general, to help the reader to better understand potential future developments in Brazil. And the possible role of biofuels within the paradigm energy shift. In a globalized world with so much interconnected, any impact will resonate beyond national boarders, even in Brazil with its unique conditions for biofuel development.

A fundamental question we need to pose in this shifting scenario is, *what could be the future of biofuels in a new scenario in which no fuel will have the monopoly of oil and gas? How changes in the rest of the world will affect biofuels, particularly in Brazil?* A scenario that will be far more complex, mixed, volatile, and more geographically determined. Biofuels cannot operate on a vacuum, and hence what happen in other sectors, particularly oil, will shape their development. Take, for example, electric vehicles where in some parts of the world, as Europe, could relegate biofuels. In addition, there are two key drivers that will affect the outcome of the energy sector, and consequently biofuels: international response to climate change and environmental sustainability.

1.2. GENERAL BACKGROUND

The energy sector is facing a huge transformation catapulted by growing concern with climate change, global warming, decarbonization, environment, sustainability issues, and political volatility. This is further complicated by Russia's war in Ukraine whose final outcome is difficult to predict. Never before in history has there been so much social and political pressure to change.

The oil sector, and more specifically transport, has not been synonymous of rapid change in the past. In fact, for decades, hardly any significant change took place. For decades, the automobile industry and oil giants, have shared a marriage of convenience. However, the undergoing changes are unparalleled. There are many "unknowns" but undoubtedly such changes represent a paradigm shift with cataclysmic proportions. Huge economic, social, environmental and policy implications, with hundreds of billions of dollars at stage. Such changes represent both huge opportunities and also serious challenges and pitfalls. There is also huge uncertainty with regard to climate change whose enormous implications are difficult to predict; such outcome will have a major global impact, determining future directions and hence will be a key driver for change.

There will be a long energy transition, difficult to predict. But with the right policies, change can be like "a snowball" once started will be difficult to stop. It could work by its own inertia. Given the enormous resources potentially available, the world capacity for scientific and technological developments and innovations, such transition could be faster than predicted.[1] For example, climate warming has been ignored for far too long but now that the world community is beginning to take it more seriously, and social, and political pressures will increase to force change. Climate change is beginning to sink into the young population psychic, despite the fact that for large part of the world population it is a daily struggle for survival and is unwilling or unable to make the necessary sacrifices today for such changes, perceived as something that may happen in a long-distance future.

Within this context, what could be the potential role of biofuels in transport? Biofuels have been around for many decades, and in particular in Brazil and USA, and to a lesser extent in many other countries, hence they represent little technological challenge by comparison, as it is a well-known and proven technology and fuel. The pros and cons of biofuels have been heatedly debated, but on balance, if produced and used sustainably, the benefits far aweigh the negative impacts. Therefore, biofuels represent a considerable advantage over many other emerging alternatives in many parts of the world.

Biofuels have also been the object, as most renewables, of booms and busts. When such alternatives appeared to be economically viable, the price of oil came down dramatically e.g., early 1990s. This time would be different because pressure from climate warming and the environment. Hence, growing concern with climate change, environment, and sustainability, together with social and political pressures, means that this paradigm shift cannot be reversed. On the contrary, it is expected to be accelerated.

But biofuels also have also their negative side that limit their potential. The main one being land competition with food production and the requirement that biofuels need to be environmentally sustainable. The "food versus fuels" (FvF) dilemma has

1 The amount of data generated daily, and the speed in which new scientific data is generated, is such that it is impossible to keep pace except for very specific topics. This poses serious problems.

been the object of heated debate for decades. The debate (e.g., see ROSILLO-CALLE & JOHNSON, 2010; ROSILLO-CALLE, 2018) has been clouded by vested interest, bias, and lack of reliable scientific data. The "FvF" debate needs to consider the intertwined nature of biofuels, land use, food production, food waste, animal feed production, diets, social habits, food distribution systems, policy, and so forth. Advanced biofuels (G2 and G3) could avoid most these problems. It is generally accepted that food and biofuels production can be complementary and, in any case, food production will always prevail.

Biofuels have been around for decades and are an integral part of the fuel mix in the transportation systems. Currently this sector provides between 3-4% of fuel consumption in road transportation, with USA and Brazil as the main producers and consumers, and to a lesser scale in the EU, China, and India (see Figure 1.4). It is an unequal and uneven contribution on a global scale, representing a minuscule contribution in many countries, or none at all. This uneven distribution reflects resource endowment, cultural factors, know-how and policy options. The totality of biofuels is used blended with petrol and diesel in varying proportions, except in Brazil where about 50% is E100.

Another question we need to address is the extent to which the paradigm shift will enhance or inhibit their development, as oil is gradually replaced by other emerging alternatives, e.g., electricity, hydrogen, solar, etc. However, new technological developments such as G2 and G3, e.g., converting cellulose-based biomass to ethanol, will make available large additional amounts of feedstocks in many more countries, at competitive cost over time. Also, we need to ask: should biofuels be promoted as a vehicle fuel or should the focus be on new emerging markets e.g., aviation, maritime transport, or fine chemicals? Biofuels will not be the answer to oil, but they can play a much greater role than has been the case so far in the fuel mix and can make an important contribution to the solution.

During the oil crisis of the 1970s, Brazil showed a great vision by setting up a national ethanol fuel program (e.g., see ROTHMAN, GREENSHIELDS & ROSILLO-CALLE, 1983), a unique project at national level, that served as an international school and was, at the same time, the envy of many countries. This initiative put Brazil in a unique historical footage e.g., a considerable know-how on alternative fuel for transport, as well as technological, agricultural, economic, environmental, and social benefits. Thus, a key question we need to ask is: what will be the new role of biofuels in the emerging energy paradigm in Brazil? What lessons can be applied from its unique historical experience? What lessons are there for other countries? Such questions are addressed in the following chapters.

1.3. ENERGY/TECHNOLOGY TRENDS

Perhaps never in history has there been so much investment and research aimed at finding new alternatives to oil and gas, not only new fuels such as hydrogen, but also engine modifications (electric vehicles) and battery technology. Given the increasing amount of resources being allocated, technological innovations are growing spontaneously e.g., vehicle batteries, wind power, solar power, hydrogen, new feedstock, to name a few.

It can be stated that renewable energy (RE) in general is growing rapidly, though not necessarily at the pace of potential demand (this will require a fourfold increase), and very unequal around the world. There is a growing shift toward Asia (e.g., China and India), which could overtake the EU. Table 1.1 summarizes the main trends around the world in 2021.

Table 1.1 Summary of main global renewable energy (RE) indicators 2019-2020. Source: REN21 (2022).

Year		2019	2020
Investment on renewables (RE) (power & fuels	Billion $US	298.4	303.5
RE power capacity, including hydro	GW	2,581	2,838
RE capacity (excluding hydro)	GW	1,430	1,668
Hydropower capacity	GW	1,150	1,170
Solar PV capacity	GW	621	760
Wind power capacity	GW	650	743
Bio-power capacity	GW	137	145
Geothermal power capacity	GW	14.0	14.1
Concentrating solar thermal power	GW	6.1	6.2
Green power capacity	GW	0.5	0.5
Ethanol production (annual	Billion litres	115	105
FAME biodiesel production (annual)	Billion litres	41	39
HVO biodiesel production (annual)	Billion litres	6.5	7.5
Countries with biofuels blend mandates	–	65	65

Note: for further details and footnotes see source, page 13. *Please note there are some anomalies due to the Covid-19 pandemic.*

Among the indicators, it calls the attention the total investment going to RE, a firm indication of how investors are taking seriously RE. This trend is increasing rapidly due to recent political considerations such as growing concern with climate change and the Russia-Ukraine war. It demonstrates that RE is reaching maturity.

Annual demand for biofuels (bioethanol and biodiesel in particular) is expected to increase (see below) to around 190 billion litres (B/l) by 2023. Biofuel's production is (and will continue) to be very unequal around the world. Large programs are unlikely, with many small to medium side programs, scattered among many countries, and use primarily as blends. This panorama is unlikely to change fundamentally, with USA and Brazil continuing to dominate this market. It is difficult to predict how this new energy transition will be shaped, as policies are very fragmented around the world together with volatile geopolitics, reflecting national priorities, cultural and economic factors, resource endowment and know-how.

Two new candidates seem to be particularly strong, i) electrification and ii) hydrogen. None of them are really new as they have been around for a long time (see below). What is new is the huge amount of resources and R&D going into these sectors. Electrification of transport is a real challenge and poses serious problems e.g., how to ensure electricity is generated from RE, and problems posed by such large demand for batteries. In fact, there is a serious danger of creating a monopoly, as with oil, in some countries by relaying too much on electricity. Given the nature and control of rare material by just a few countries, pursuing this option is problematic. There are, therefore, serious economic, political, and environmental unsolved and unanswered problems.

Costs of these alternatives are coming down quite considerably and becoming competitive with conventional fuels, at least in the medium term. Despite the promise of such alternatives, it seems conventional RE (hydro, solar, geothermal and biofuels) will continue to grow rapidly and being a major alternative (IEA, 2021a).

1.4. ENERGY SCENARIOS

There have been many energy scenario predictions but hardly any of them have turned out to be correct. There are a multitude of possible scenarios, but despite improvement of data, energy uncertainly remains an integral part of the world's volatile geopolitics, and a multitude of changing political and economic factors.

One thing seems certain, the automobile of the future will be very different, in which decarbonization, environment and sustainable fuels will play a key role. Such changes represent an enormous challenge to both the automobile and oil industries e.g., technological, environmental, economic, and social. While oil is not going away any time soon it will, progressively, be phased out over time. And many candidates are emerging e.g., electricity, hydrogen, solar, modern biofuels (bioethanol, biodiesel). The future will be dominated by a mixture of fuels with none of them playing a dominant role as oil.

Figure 1.1 shows the changing nature of energy markets, 1900 to 2050. As can be appreciated all RE experience a rapid increase from 2020 and become the main source of energy, with nearly 60%, while fossil fuels declined considerably in the Accelerated Scenario (see below). Decarbonization plays a major role in the increase of RE and a sharp decline of fossil fuels.

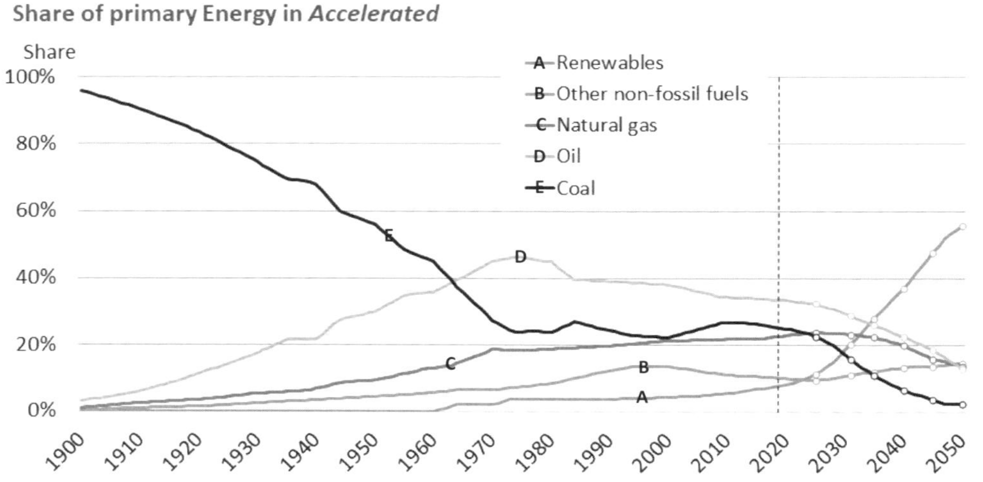

Figure 1.1 Changing nature of global markets: more diverse energy mix, increased competition, and greater customer choice. Source: BP (2022).

The following figure (Figure 1.2) shows in more specific details the role of modern bioenergy according to BP three scenarios: Accelerated, Net Zero and New Momentum, from 2019 to 2050.

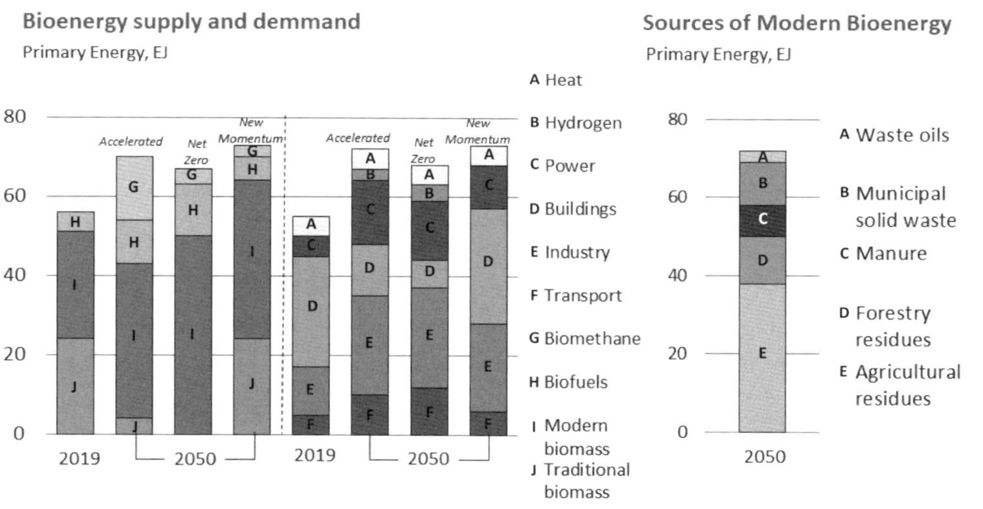

Figure 1.2 Modern bioenergy increases sharply, supporting the transition to a low-carbon energy system. Source: BP (2022).

BP has developed three main energy scenarios: *Accelerated Scenario, Net Zero and New Momentum.*[2] The first two are broadly in line with the Paris Agreement (UN

2 For a detailed assessment of the three energy scenarios, please see BP (2022) Energy Outlook 2022 Edition (www.bp.com/energy/global)

COP25). This requires maintaining temperature below 2°C. Global energy demand measured at the final point of use (total final consumption - TFC) peaks in all scenarios as gains in energy efficiency accelerates. TFC peaks in the early 2020s in the Net Zero, around 2030 in the Accelerated and in mid-2040 in the New Momentum (BP, 2022, p. 17).

Total final energy consumption across all emerging economies growth by about 35% and 5% in the New Momentum and Accelerated and falls by 20% in the Net Zero. In the developed world, contrary, the TFC falls by 25-50% in all three scenarios by 2050s. The share of fossil fuels in TFC declines from around 65% in 2019 to 30-50% in all three scenarios, with the greatest fall corresponding to coal (BP, 2022, p. 39).

It is important to keep in mind that scenarios are largely a function of the level of decarbonization and a multitude of other factors. For details of world energy consumption and projections see also the International Energy Agency (IEA) e.g., IEA (2021a); IEA (2021b); IEA (2021c). There are many others excellent reports and databases on energy e.g. (https://ourworldindata.org/energy-key-charts) where you will find a wealth of excellent energy data.[3]

1.5. BIOFUEL SCENARIOS {CONVENTIONAL AND ADVANCED (G2 AND G3) BIOFUELS}

Figure 1.3 summarizes final use of biofuels demand, based on the IEA (2017) data. The main striking point is the rapid growth of advanced biodiesel and biojet. It is important to state that it is one of many scenarios and hence should be seen as just a *potential outcome*. Scenarios are just potential projections, but despite the discrepancies, all of them point to the same direction: a rapid increase in RE and a steady decline of fossil fuels.

The reason why this Figure 1.3 has been chosen is because data for the latest few years have been distorted by the Covid-19 pandemic. There are three new factors not fully recognized in this figure: i) the rapid growth of electricity and hydrogen, ii) much greater concern with climate change, and iii) growing energy uncertainly and volitivity. This new political reality (caused by Russia's war on Ukraine, among others) could potentially speed up the development of all RE considerably or, contrary, delate it implementation. For example, India is increasing the blending of ethanol with gasoline, primarily to save oil imports due to rising oil prices.[4] And in Europe, which is highly dependent on Russia fossil fuels, politicians are struggling to reduce such dependency with a multitude of options under consideration. Time will tell.

---

3 See also IEA (2021) Key World Energy Statistics (www.iea.org), IEA (2021) World Energy Statistics and Balances 2021 (www.iea.org)

4 See for example The Times of India, 19th May 2022. India to launch20% ethanol-mixed gasoline (...) (www.thetimesofindia.com)

The evolution of such fuels is strongly dependent on environmental, and climate change and political considerations. If concern with climate warming accelerates significantly, then such scenarios could be quite realistic. Huge resources are being channelled towards this end.

The main biofuels (conventional and modern) continue to be bioethanol and biodiesel. The expected growing role of biojet is somehow a bit controversial as other models give a much lower potential contribution. Much depends on growing demand of such sectors such as aviation and maritime transport. (IEA, 2017). See Section 10.

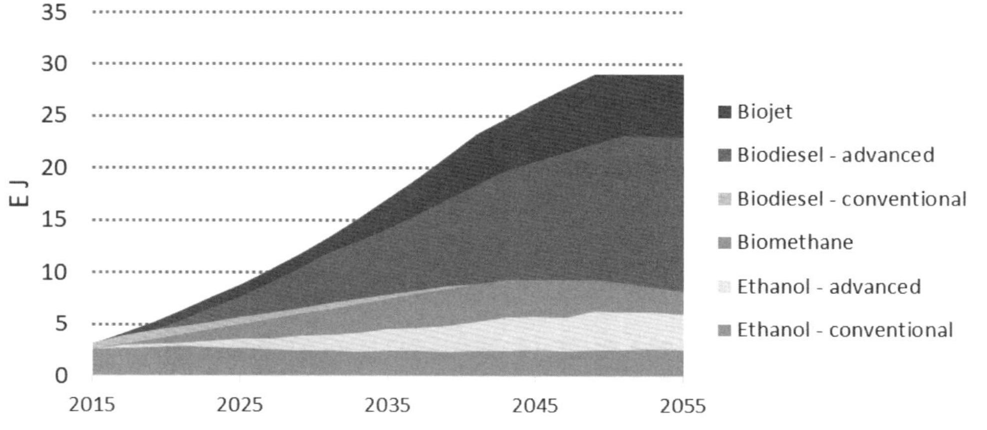

Notes: Conventional biodiesel referes to crop based FAME biodiesel; Advanced biodiesel refers to a range of advanced biofuels suitable for use in the diesel pool.

Figure 1.3 Biofuels final transport energy demand by fuel type in the IEA 2DS, 2060. Source: IEA (2017).

As for the type of fuel and geographical distribution, the USA and Brazil continue to dominate biofuel production and consumption, with Europe remaining more or less stable and small increases in Asian markets. A notable increase is advanced biodiesel and biojet fuel. Indonesia is set to be an important player on biodiesel, although it is environmentally controversial because such expansion is based on palm oil plantations (see Figure 1.6).

1.6. ETHANOL FUEL

The following figure (Figure 1.4) shows the geographical distribution of ethanol fuel, and global production in 2021. As can be appreciated, the USA and Brazil are the key players, followed by the EU and some Asian countries (China, India, Thailand). The rest of the world represent a very small market, and in many none at all.

Ethanol was the first candidate as vehicle fuel rather than gasoline. And in Brazil as early as 1925 there was a Federal Law that made compulsory blending ethanol with gasoline (e.g., see ROTHMAN, GREENSHIELDS & ROSILLO-CALLE, 1983); see also Chapter 6).

Ethanol has been the king as transport fuel around the world. It has been growing steadily, except for 2020 due to the Covid-19 pandemic. It is not expected to expand exponentially as a few countries have the capacity to increase production in large scale and even in Brazil there are questions as its future expansion (see Chapter 5). Total world production of ethanol fuel in 2021 was estimated at 102.2 B/l.

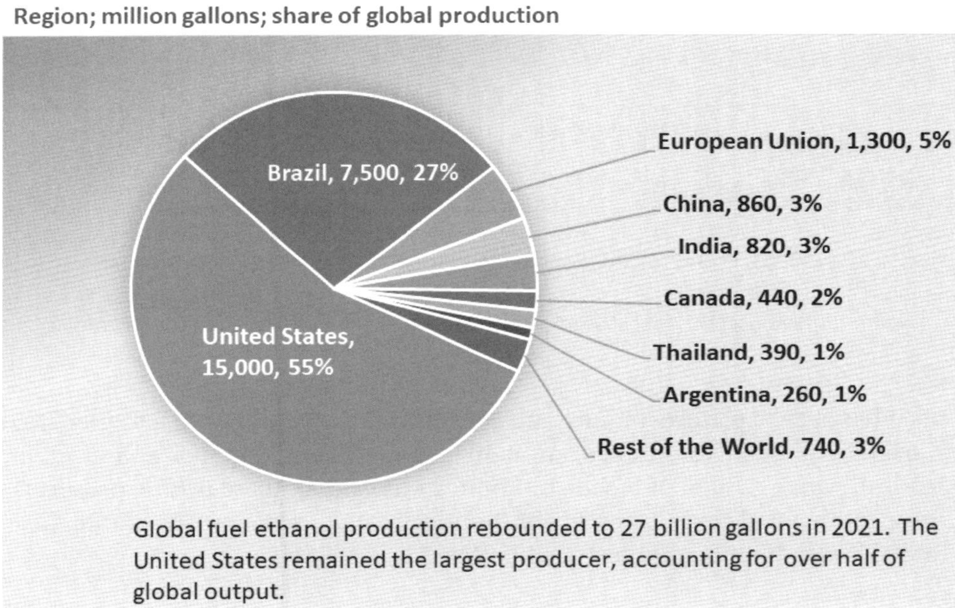

Figure 1.4 Global Fuel Ethanol Production in 2021. Source: RFA (2022).

Note: One USA gallon = 3.785 litres

The USA is the world largest ethanol producer and consumer, representing 55%, followed by Brazil with 27%. Figure 1.5 shows the historical production of ethanol in the USA, rising from almost zero in 1981 to about 56.70 B/l (c.15 B/gal) in 2021. Despite this impressive achievement, current policy indicates the USA does not plan to increase ethanol production in any significant scale in the near future. Unlike USA, Brazil has considerable potential to increase both production and consumption significantly, depending on governmental policy, as discussed in detail in the following chapters.

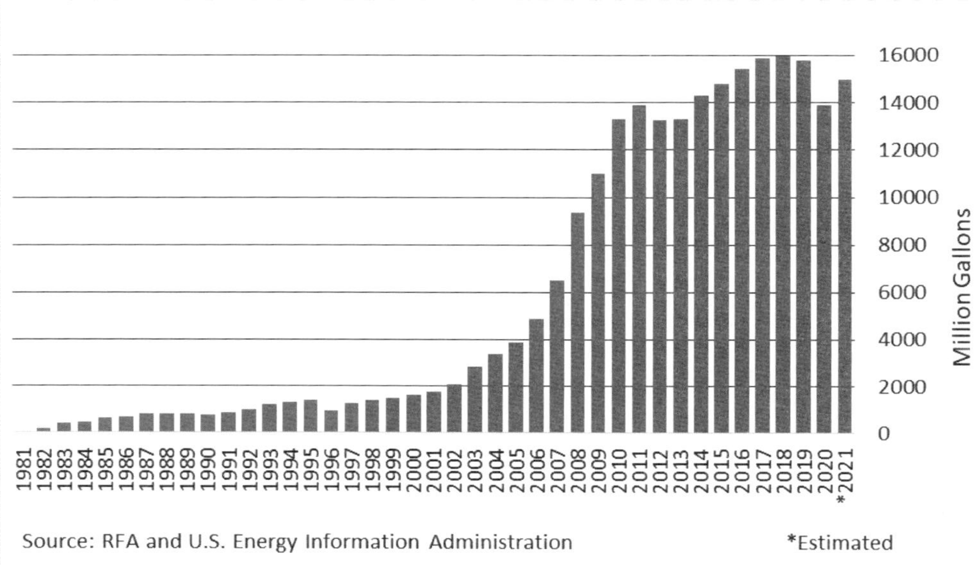

Figure 1.5 Historical U.S. fuel ethanol production. U.S. Energy Information Administration. Source: RFA (2022).

As it has been explained, biofuels are globally unequally produced and used. Many countries are unlikely to produce any at all. A good example is Africa (Sub-Saharan countries), where despite of having a significant potential, there is little prospects in the near future for a combination of political, social, cultural factor, in addition to lack of capital, know-how, etc, as illustrated in Box 1.1.

BOX 1.1: THE ROLE OF BIOETHANOL IN MOZAMBIQUE´S ENERGY TRANSITION BY RUI DA MAIA[5]

The Republic of Mozambique is a country with a vast territory of about 800 000 km2, colonized by the Portuguese, as Brazil, and with a population of around 30 million. Although, there are fundamental differences between the two countries, both can be considered medium-to-low income, where commodities are responsible for a significant share of exports.

Imports consist of fuels, machinery, and agricultural products, representing about 50% of country´s imports. South Africa, China, Emirates, and The Netherlands are

5 Professor at Universidade Técnica de Moçambique (UDM), Mozambique, damaiarui59@gmail.com

the most important partners.[6] *The country exports fuels and metals represent nearly 73%, and India, South Africa, and The Netherlands are the most important trade partners. It can be added that South Africa and India are Mozambique´s traditional partners, while China is gradually increasing its importance.*

In the energy area, Mozambique has large oil reserves and petroleum plays an important part in its economy. The country has four oil refineries: S&S Refinery, CPF, South Oil, and Southern Refineries (Maëva). Although with important oil reserves and refineries, the country suffers energy supply difficulties particularly in the rural areas due to lack of electricity and other transport fuels.

As for agricultural, the country has a great potential, with fertile land and abundant water, although the country still needs to import basic products. The main reason for the food insecurity of Mozambique relies in the precarious agrarian system where land tenure is the main issue. The present agrarian system can be considered a major obstacle to a more organize and modern agricultural production system, either for food or biofuels production. On top of difficulties associated with land, there is chronic lack of capital, and skills.

The Civil War that followed the Mozambican independence destroyed much of the country´s agriculture and rural infrastructure. The colonial period had its negative consequences, associated with the problems related with Civil War. Although the country has had relatively high growth rates after 1994, this was not sufficient to compensate and rebuilt what was destroyed.

Also, when government authorities considered a biofuel program in the 2000s, crops such as Jatrofa Curcas were considered. As it will be discussed later in this book, the adoption of a non-energy crop for biofuel production is not recommended. Mozambique has experience with sugarcane and should try to use it in any biofuel program.

Sugarcane sector in Mozambique is quite large and organized. The annual sugarcane production is around 3 million tons of cane/year.[7] *The Mozambique Sugar Producers Association (APAMO) counts with 4 mills producing around 530,000 tons of sugar and an estimated area of 57,900 ha, directly employing more than 140,000 people.*[8] *These mills are owned by South African companies.*

6 http://portalcomercioexterno.gov.mz/pt/trade-data#:~:text=Os%20principais%20grupos%20de%20bens%20importados%20foram%20os%20Combust%C3%ADveis%20Minerais,USD%20872.6%20milh%C3%B5es%20(12.2%25).

7 FAO (2018)

8 https://i-enterpriser.com/2022/02/27/governo-investe-no-fomento-da-producao-de-cana-de-acucar/

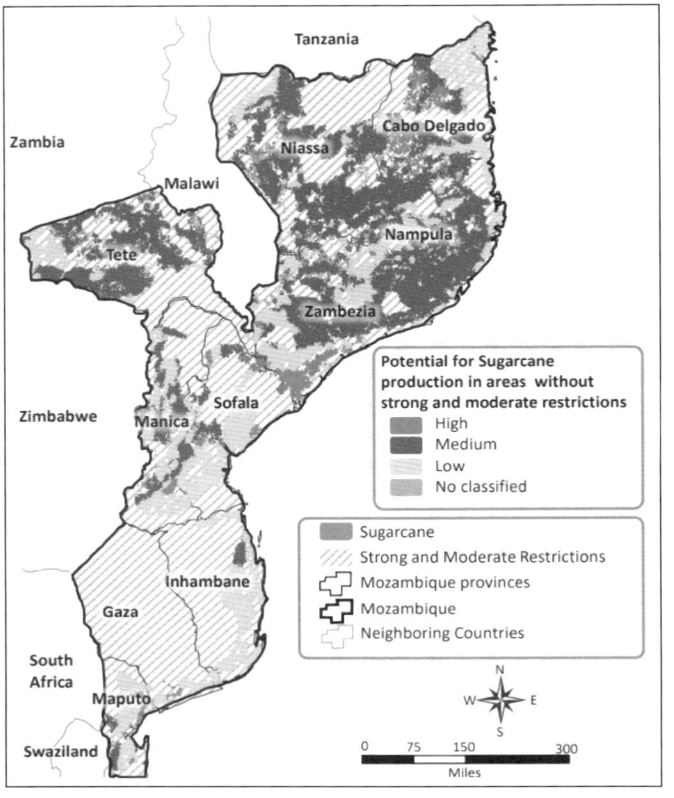

Source: Moreira et al. (2019)

Regarding fuel sugarcane ethanol production, in 2009 it attracted attention from Brazilian investors (UNICA). Basically, three projects were considered, one in the south (120 million litres/year), one in the central region – Manica (100 million litres/year), and one in the North (300 million litres/year). Investments in sugar and ethanol are considered strategic in Mozambique because former colonies are allowed to export to European countries without import taxes.[9] *A study conducted by MOREIRA et al. (2019) estimates that 33 Mha are suitable for sugarcane cultivation in Mozambique with medium and high potentials, mostly in the North part of the country (figure below: Source: Moreira et al. (2019)).*

Naturally, the role of ethanol production in Mozambique would be more oriented to achieve socio-economic goals, such as increasing exports, population income, and jobs. Therefore, the GHG emissions could be considered a secondary achievement. Another important role of biofuels production would be the introduction of a new

9 https://www2.senado.leg.br/bdsf/bitstream/handle/id/448956/noticia.htm?sequence=1

agrarian system, in which the most important stakeholders could be included. It can be said that, without including the traditional landowners in the negotiation, hardly any one of these goals would be achieved. Therefore, sugarcane ethanol could play an important role in the energy transition in Mozambique.

Food versus biofuels is indeed very sensitive and an important consideration in Mozambique. This may explain, at least partially, FAO doubts in supporting biofuels production and hence its recommendation to use crops that do not compete directly with food production. FAO was probably more concerned with food security than with energy security or jobs, and probably thinks that non-food crops are a safer option when biofuels production is considered.

Mozambique is a country where, despite its considerable biofuel potential, is a good example in which biofuels would not have any future in the short term (5-10 years). This is due to a combination of factors e.g., lack of investment, skill, and political will. Since the social and political situation in most of Sub-Saharan countries are quite similar, one could conclude that this Continent offers little prospects for biofuels in the short term, at least in any significant scale.

Mozambique, as other middle- and low-income economies, doesn´t have means to absorb rapidly the electric mobility. High vehicle prices and lack of necessary infrastructure are the main drawbacks. This transition will probably take several decades. In the meantime, several technologies/fuels will coexist, creating additional pressure in infrastructure and services. Most likely, multinational corporations from China or Europe will start to introduce electric vehicles coexisting with internal combustion engines.

REFERENCES

Moreira, M. M. R.; Gomes, F. H.; Costa, K. M. Case study: Potential of sugarcane production for Mozambique. In Sugarcane Bioenergy for Sustainable Development. *Routledge, London, 2019, 417 p..*

Table 1.2 summarizes the major feedstocks used in the production of conventional biofuels. For ethanol, the dominant sources are sugarcane in Brazil and maize in the USA. For biodiesel are mainly soybean (Brazil and USA) But feedstocks are becoming more diversified and with the advance of biofuels (G2 and G3) using woody biomass, and oil-bearing plants will be of much greater importance.

Table 1.2 Biofuel production ranking and key feedstocks. Source: OECD/FAO (2019), "OECD-FAO Agricultural Outlook", OECD Agriculture statistics (database)[10]

Production ranking (base period)			Major feedstocks	
	Ethanol	Biodiesel	Ethanol	Biodiesel
United States	1(50%)	2(19%)	Maize	Soybean oil/diverse other oils
European Union	4(5%)	1(36%)	Maize/wheat/sugar beet	Rapeseed oil/wastes oils
Brazil	2(24%)	3(12%)	Sugarcane	Soybean oil
China	3(8%)	8(3%)	Maize	Wastes oils
India	5(2%)	15(0.5%)	Molasses	Palm oil
Canada	6(1.6%)	10(1.4%)	Maize	Wastes oils
Indonesia	23(0.2%)	4(10%)	Molasses	Palm oil
Argentina	9(1%)	5(7%)	Maize/sugarcane	Soybean oil
Thailand	7(1.5%)	6(4%)	Molasses/cassava	Palm oil
Colombia	13(0.4%)	9(1.5%)	Sugarcane	Palm oil
Paraguay	15(0.3%)	19(0.03%)	Maize/sugarcane	Soybean oil

Note: Percentage numbers refers to the production share of countries in the base period.

1.7. BIODIESEL

Large scale biodiesel production has been limited by the cost of feedstocks (e.g., soybean, palm oil) because, contrary to sugarcane and corn, productivity is much lower and costs much higher e.g., feedstock can represent 70-80% of the total costs. The markets are also quite different and fragmented. None of the major ethanol producers and consumers (Brazil and USA) have been at the forefront of the biodiesel market, partly for limitations of the feedstock. The most important market has been the EU where, contrary to ethanol, biodiesel has received greater priority. Figure 1.6 summarizes the main producers of biodiesel. The EU remains the main player followed by the USA, Indonesia, and Brazil. As can be appreciated, this market is not expected to growth globally in any large scale. USA and Indonesia are expected to grow faster, but much will depend on anergy policy developments, and the availability of large-scale sustainable feedstocks.

As said, and contrary to ethanol, biodiesel production in the USA may increase significantly. In January 2021, the county produced 2.4 B/gal (just over 9 B/l), in 75 biodiesel plants (www.eia.todayinenergy/details) (EIA, 2022), which is still small in comparison to fuel ethanol.

10 http://dx.doi.org/10,1787/agr-out-data-en

Biodiesel is, and will remain, a much smaller market than bioethanol for transportation for reasons stated above. There are currently six major global markets for biodiesel, EU, USA, Indonesia, Brazil, and Argentina and to a lesser extent, Colombia. No major changes in any these markets are expected except, perhaps, in Brazil where production could increase significantly, and Colombia that has already a growing and dynamic market, be still in a small scale. There are other fuels, e.g., biogas and biomethane, which are also a growing rapidly. Brazil, in particular, has considerable potential e.g., by use of stillage from ethanol sector (see Chapter 4).

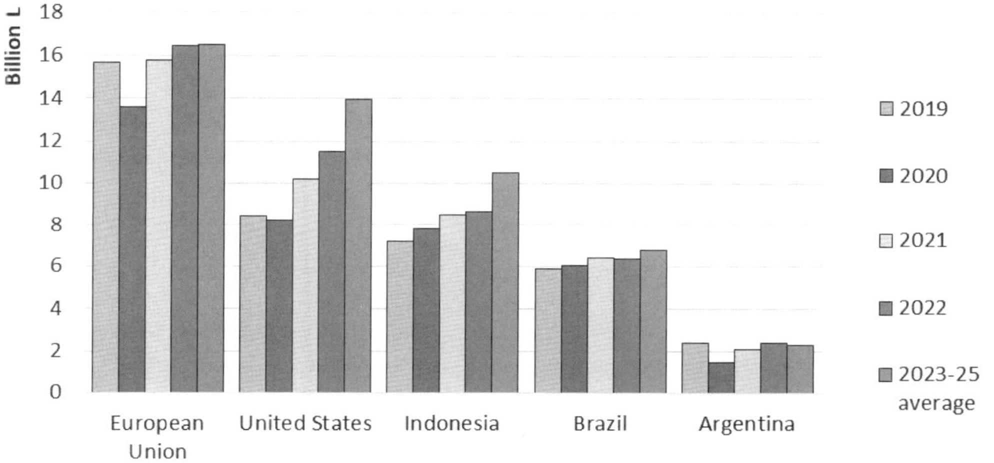

Figure 1.6 Biodiesel and HVO Production and Key Markets, 2019 – 2025. Source: IEA (2021c).

1.8. ADVANCED BIOFUELS (G2 AND G3)

There is a huge expectation as to the future potential of advance biofuels. Up to now, there have been far more failures than successes. The main challenge is reducing costs and the sustainability of the feedstocks in large scale. There is an astonishing array of feedstocks, and end-products under consideration. Unfortunately, few will succeed. The future of bioethanol and biodiesel are strongly linked to such advances, together with environmental considerations. But because environmental and energy security considerations, the expectations are high e.g., some estimates put the market growth of about 46% annually and be worth close to $50 billion within this decade (www.technavio.com).

Figure 1.7 shows global operational capacity of advanced biofuels, up to 2025. As can be appreciated, hydrotreatment of feedstocks is the most important alternative under consideration. Brazil is one of the countries with the largest potential (see Chapter 5).

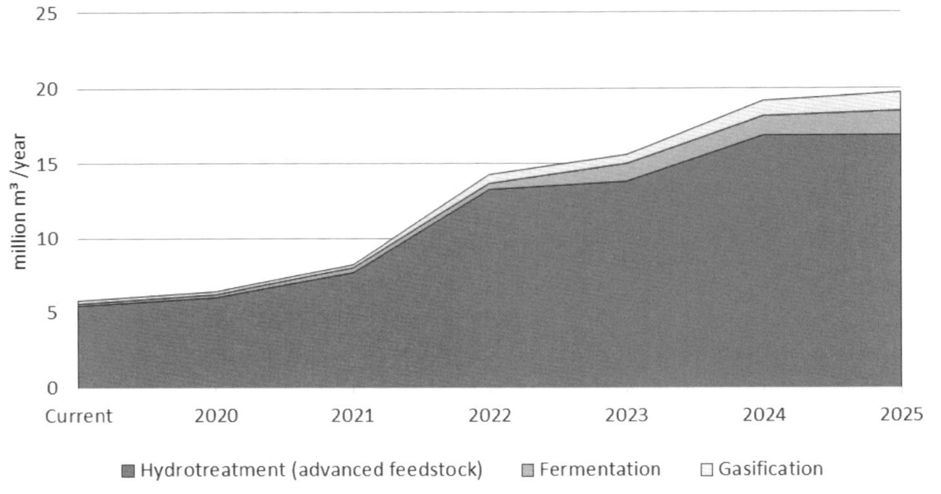

Figure 1.7 Global production capacity of advanced biofuels (operational). Source: CIT (2020).

The potential contribution and expectations on advanced biofuels vary considerably around the world. Take for example Southern Africa, as illustrated in Box 1.2, which examine the pros and cons of such potential developments. According to Gorgens, the expansion of G2 and G3 biofuels in Southern Africa would only be possible in combination with the development other RE.

BOX 1.2: EMERGING OPPORTUNITIES AND APPROACHES FOR THE PRODUCTION OF ADVANCED BIOFUELS IN SOUTHERN AFRICA BY JOHANN GÖRGENS[11]

INTRODUCTION

The industrial production of biofuels for transportation in the southern Africa context remains in its infancy, both for first-generation biofuels and more advanced options. Although approximately 400 million litres of bioethanol are produced in the region annually, the majority of it is used as potable or beverage grade products (Deenanath et al., 2012). The local biodiesel industry also remains small, and largely constrained by superior prices for waste vegetable oils, offered by European importers (Chiaramonti and Goumas, 2019). Considering the limitations on local government support for the implementation of biofuels production, the drivers for these remain as the ambitious climate change and carbon neutrality targets of the private sector, where biofuels have to compete against other renewable alternatives, and the export markets. The reduced attractiveness

11 *Professor at the Department of Process Engineering, Stellenbosch University, South Africa, jgorgens@sun.ac.za*

of crop-based biofuels to the export markets is driving a movement towards waste-based biofuels (Doumax-Tagliavini and Sarasa,2018), where industrial technologies remain to be demonstrated. In the context of both ambitious climate change targets and export markets, future opportunities for development of the biofuels sector in southern Africa seem to be focussed on second- and third-generation feedstocks and processing technologies, which offer more attractive opportunities in both regards.

SECOND- AND THIRD-GENERATION FEEDSTOCKS FOR BIOFUELS (G2 AND G3)

One of the primary examples of waste-based lignocelluloses potential available for conversion to advanced biofuels are various types of harvesting and bioprocessing residues, such as corn stover, forestry residues, sugarcane harvesting residues [SAEON, 2022]. There are, however, limitations to what portion of these lignocelluloses can be applied as feedstocks for conversion, as excess removal from the land may reduce soil fertility and productivity [Cherubin et al., 2019; Rocha et al., 2018).

Residues from industrial biomass processing such as paper waste sludges and various types of food processing wastes, which are presently landfilled, may offer the most attractive G2 resources, also considering substantial environmental and financial benefits with the avoided disposal to landfill of high-moisture organic wastes. Conversion methods such as anaerobic digestion (biogas product upgraded to compressed or liquefied methane) (Tratzi et al., 2022), fermentation (to ethanol) (Mahmoodi et al., 2018) and gasification (to produce syngas) (Chan et al., 2019) may be applied to organic wastes destined for landfilling, offering substantial environmental and economic benefits.

Several of the agroprocessing residues such as sugarcane bagasse are utilised in industrial facilities as energy feedstocks; liberating a portion of these for further valorisation can be achieved by substantial investments to increase energy efficiency in these facilities (Birru et al., 2019). Conversion of sugarcane lignocelluloses such as bagasse into alcohols through integrated first- and second-generation biorefineries, in so-called 1G2G facilities, holds substantial potential to reduce the costs of lignocelluloses conversion, through process integration and economies of scale benefits (Gnansounou et al., 2015). The 1G-portion of the resulting 1G2G biofuel would thus be classified as crop-based, while the 2G-portion would be waste-based.

Invasive alien plants (IAPs) and bush encroachment represent two largely underutilised biomass feedstocks in the region (Vera et al., 2022). Regular clearing activities are required to manage and eradicate these biomasses, often with spill-over benefits in groundwater flows and supplies to dams. Although some programs for their clearing/eradication may seem to offer access to low-cost feedstocks for valorisation, the reality is not much higher investments in these activities are required, which ultimately must be paid for by the biofuels production facility. This results in these feedstocks being as costly as other types of lignocelluloses available for processing.

Advances in the development and commercialisation of CO-gas fermentation to ethanol have availed new opportunities for third-generation biofuels production from industrial offgasses (Petersen et al., 2021). Several of the minerals processing smelters in the South African mining industry produce CO-rich offgasses, which have substantial climate change impacts. Capturing, compression and fermentation of these gasses results in the production of ethanol.

LIGNOCELLULOSES PROCESSING VIA GASIFICATION-SYNTHESIS

The processing of lignocellulosic biomass through gasification-synthesis processes is an attractive opportunity for local implementation, due to South Africa being a world-leader in commercial implementation of gasification and Fischer-Tropsch (FT) synthesis technologies, for the production of gasoline, diesel, jetfuel and a range of chemicals from coal (Gupta et al., 2020). These industrial facilities offer opportunities to reduce greenhouse gas emissions by partial or complete replacement of coal with natural gas and/or lignocellulosic biomass. Existing industrial facilities may contribute to biomass-conversion into biofuels through either co-gasification of biomass with coal in existing gasifiers (Quintero-Coronel et al., 2022), or by erecting biomass-only gasifiers and co-feeding the resulting gas product with syngas derived from coal or natural gas. Unique technical and economic opportunities also exist for co-processing/reforming of natural gas with biomass-derived syngas (Gardezi et al., 2013), which may be explored further in future. Both of these approaches will rely on the allocation of the renewable carbons, co-fed into synthesis steps with fossil-derived carbon, to specific products of interest, through the so-called mass-balance approach (Rosenfeld et al., 2020). This will allow for the recovery of the increased cost of biomass compared to fossil fuels from specific products that are able to attract premium prices in the market.

Alternatively, biomass conversion through gasification-synthesis may be pursued through dedicated, greenfields, biomass-only processing facilities, which may either be decentralised or centralised, depending on the locations of biomass resources. Such facilities will require that all products (gasoline, diesel, chemicals, jetfuel) be sold at premium prices, to recover the costs of biomass procurement (Petersen et al. 2022). Recent developments in FT-synthesis have therefore targeted new catalysts able to increase the proportion of sustainable aviation fuel (SAF) produced from such facilities, to maximise the commercial potential of these facilities.

ENVIRONMENTAL CONSIDERATIONS FOR SECOND- AND THIRD-GENERATION BIOFUELS

The implementation of second- and third-generation biofuels production is largely driven by expected environmental benefits, either for local industries or the export markets. In this regard the allocation of environmental burdens to a range of products typically obtained from biomass-processing facilities, warrant careful consideration. Together with the "greening" of utilities for processing/conversion of biomasses, these two factors have substantial impacts on the environmental credits attributed to final biofuel products.

As a first example, sugarcane mills that produce biofuels should preferably continue to co-produce sugar with biofuels. This will allow a substantial portion of the environmental burdens associated with sugarcane cultivation, harvesting and processing to be allocated to the sugar product (Maga et al., 2019), which often faces fewer market-acceptance hurdles due to environmental impacts such as greenhouse gas emissions. This situation may well change in future where food markets are gradually becoming more sensitive to the climate impacts of food products.

On the other hand, the processing of industrial, CO-rich offgasses into bioethanol through gas fermentation, requires substantial amounts of electrical energy for gas compression (Petersen et al., 2021). Such electricity should be provided from renewable energy sources such as wind or solar, rather than being produced from a portion of the available CO-gas, so as to increase the amount of ethanol produced per unit of CO-feed and benefit from the resulting economies of scale.

Gasification-synthesis processes require both renewable electricity and green hydrogen obtained from substantial solar and wind resources, as co-feeds with biomass, to maximise the yields of biofuels on available biomasses, and benefit from economies of scale (Dossow et al., 2021). Whether co-feeding of these types of renewable energy feedstocks will provide immediate economic benefits, depends on their production costs, where solar and wind-based electricity is often cheaper than biomass-derived. In it clear that in future scenarios the supply of renewable and sustainable biomass not only for biofuels production, but also for a wide range of other biomass-derived products, will be a limited resource, further justifying the use of solar and wind resources to provide "green" utilities for biomass conversion.

FUTURE PERSPECTIVES

The potential expansion of biofuels production in the southern Africa context should be viewed together with simultaneous development of substantial solar and wind electricity resources, and the associated green hydrogen options, as well as electrification of transportation systems. Large investments in renewable electricity supply remain to be ramped up substantially, to provide grid-based charging options for electric vehicles, which are still substantially more expensive than internal combustion vehicles, while solar and wind-based hydrogen remains largely unaffordable to local markets. Biofuels produced from low-cost or negative cost feedstocks (such as organic wastes destined for landfilling) may offer meaningful investment opportunities for local implementation in the short-term, while in a longer horizon there will be increasing competition from other potential applications of the limited supply of renewable carbon that the region can offer.

REFERENCES

Birru E, C Erlich, A Martin (2019). Energy performance comparisons and enhancements in the sugar cane industry. Biomass Conversion and Biorefinery 9: 267–282

Chan WP, A Veksha, J Lei, W-D Oh, X Dou, A Giannis, G Lisak, T-T Lim (2019). A hot syngas purification system integrated with downdraft gasification of municipal solid waste. Applied Energy 237: 227-240

Cherubin RC, IP Lisboa, AGB Silva, LL Varanda, RO Bordonal, JLN Carvalho, R Otto, PS Pavinato, A Soltangheisi and CEP Cerri (2019). Sugarcane Straw Removal: Implications to Soil Fertility and Fertilizer Demand in Brazil. BioEnergy Research 12: 888–900

Chiaramonti D, T Goumas (2019). Impacts on industrial-scale market deployment of advanced biofuels and recycled carbon fuels from the EU Renewable Energy Directive II. Applied Energy 251: 113351

Deenanath ED, S Iyuke and K Rumbold (2012). The Bioethanol Industry in Sub-Saharan Africa: History, Challenges, and Prospects. Journal of Biomedicine and Biotechnology 2012: 416491. doi:10.1155/2012/416491

Dossow M, V Dieterich, A Hanel, H Spliethoff, S Fendt (2021). Improving carbon efficiency for an advanced Biomass-to-Liquid process using hydrogen and oxygen from electrolysis. Renewable and Sustainable Energy Reviews 152: 111670

Doumax-Tagliavini V, TC Sarasa (2018). Looking towards policies supporting biofuels and technological change: Evidence from France. Renewable and Sustainable Energy Reviews 94: 430-439

Gardezi SA, B Joseph, F Prado, A Barbosa (2013). Thermochemical biomass to liquid (BTL) process: Bench-scale experimental results and projected process economics of a commercial scale process. Biomass and Bioenergy 59: 168-186

Gnansounou E, P Vaskan, ER Pachón (2015). Comparative techno-economic assessment and LCA of selected integrated sugarcane-based biorefineries. Bioresource Technology 196: 364-375

Gupta PK, V Kumar, S Maity (2020). Renewable fuels from different carbonaceous feedstocks: a sustainable route through Fischer–Tropsch synthesis. Journal of Chemical Technology and Biotechnology 96: 853–868 https://doi.org/10.1002/jctb.6644

Maga D, N Thonemann, M Hiebel, D Sebastião, TF Lopes, C Fonseca, F Gírio (2019). Comparative life cycle assessment of first- and second-generation ethanol from sugarcane in Brazil. The International Journal of Life Cycle Assessment 24: 266–280

Mahmoodi P, K Karimia, MJ Taherzadeh (2018). Hydrothermal processing as pretreatment for efficient production of ethanol and biogas from municipal solid waste. Bioresource Technology 261: 166-175

Petersen AM, O Okoro, F Chireshe, T Moonsamy, JF Görgens (2021). Systematic cost evaluations of biological and thermochemical processes for ethanol production from biomass residues and industrial off-gases. Energy Conversion and Management 243: 114398

Petersen AM, F Chireshe, JF Gorgens, J Van Dyk (2022). Flowsheet analysis of gasification-synthesis-refining for sustainable aviation fuel production from invasive alien plants. Energy 245: 123210. https://doi.org/10.1016/j.energy.2022.123210

Quintero-Coronel DA, YA Lenis-Rodas, L Corredor, P Perreault, A Bula, A Gonzalez-Quiroga (2022) Co-gasification of biomass and coal in a top-lit updraft fixed bed gasifier: Syngas composition and its interchangeability with natural gas for combustion applications. Fuel 316: 123394

Rocha JHT, JL de Moraes Gonçalves, CB Brandani, A de Vicente Ferraz, AF Franci, ERG Marques, JCA Junior, A Hubner (2018). Forest residue removal decreases soil quality and affects wood productivity even with high rates of fertilizer application. Forest Ecology and Management 430: 188-195

Rosenfeld DC, H Böhm, J Lindorfer, M Lehner (2020). Scenario analysis of implementing a power-to-gas and biomass gasification system in an integrated steel plant: A techno-economic and environmental study. Renewable Energy 147 (1): 1511-1524

South African Environmental Observation Network-SAEON, The BioEnergy Atlas for South Africa https://bea.saeon.ac.za/ Accessed 7 July 2022

Tratzi P, M Torre, V Paolini, L Tomassetti, C Montiroli, E Manzo, F Petracchini (2022). Liquefied biomethane for heavy-duty transport in Italy: A well-to-wheels approach. Transportation Research Part D: Transport and Environment 107: 103288

Vera I, N Goosen, B Batidzirai, R Hoefnagel, F Van der Hilst (2022). Bioenergy potential from invasive alien plants: Environmental and socio-economic impacts in Eastern Cape, South Africa. Biomass and Bioenergy 158: 106340

Maga D, N Thonemann, M Hiebel, D Sebastião, TF Lopes, C Fonseca, F Gírio (2019). Comparative life cycle assessment of first- and second-generation ethanol from sugarcane in Brazil. The International Journal of Life Cycle Assessment 24: 266–280

1.9. HYDROGEN

The utilization of hydrogen as fuel is not new. In fact, it has been used for over a century but in small and specific applications, with booms and busts. In recent years, however, it has received an enormous interest and huge amount of investment is being pouring into hydrogen, especially green hydrogen. Hydrogen is very versatile fuel that can be produced from a verity of energy sources e.g., coal, oil, natural gas, biomass, other RE, nuclear, etc., through a wide variety of technologies (reforming, gasification, electrolysis, water splitting etc. (IEA, 2021e).

Global hydrogen demand in 2020 was 90 Mt, almost entirely from fossil fuels (see Figure 1.8). Seventy-nine percent was produced from dedicated hydrogen production plants (IEA, 2021e).

Green Hydrogen, produced from RE, is currently booming catapulted by concern with environment and climate change. For many analysts, this option represents the future fuel for transport. There are, however, a lot of unanswered questions since this will require huge investment, infrastructure, and natural renewable resources.... For example, if vehicle batteries were to use green electricity, the demand will be enormous.

If present projections on electric vehicles materialize, this will represent a huge change in the automobile industry. According to the IEA (IEA, 2021f), at the end of 2020 there were 10 million electric cars in the road worldwide. Europe leads with about 1.6 million, followed closely by China with 1.5 million and USA far behind with c300.000 units, which could change dramatically. All major vehicle manufactures have ambitious electrification plans. In 2020 consumers expending in electric cars reached \$120 billion, and governments around the world were expending around \$14 billion supporting electric car sales, in 2019 alone. If present trends continue, this market could reach 230 million vehicles in 2030 (IEA, 2021f).

This market is in constant a flux as technological advances cut costs and pressure to decarbonizes increases to cut pollution. This will have major consequences worldwide for biofuels. For example, in Europe with such focus on electric cars, it is unlikely that bioethanol will be a priority, unlike biodiesel which has a much brighter future. In this scenario, Brazil is a privileged position, it will have the flexibility of using bioethanol, biodiesel, and electricity, given its enormous hydro power potential (see Chapter 5). The country is already the world's largest contributor of RE in the national energy matrix.

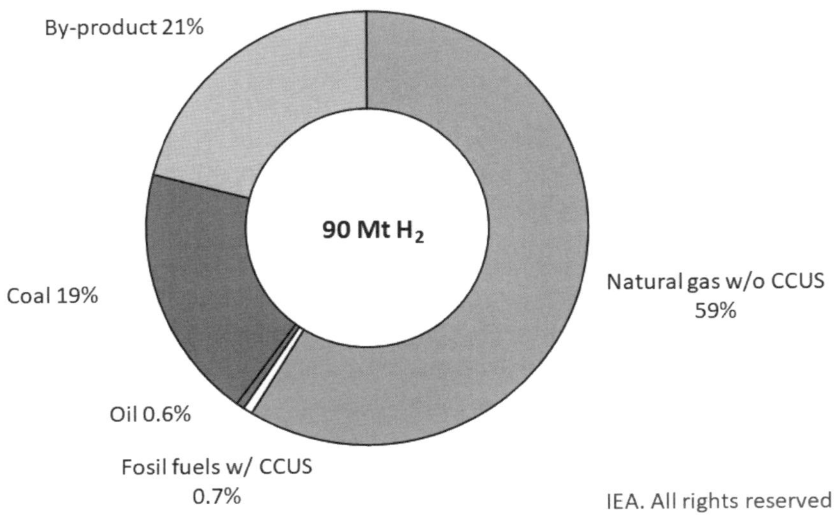

Note: CCUS = carbon capture, utilisation and storage

Figure 1.8 Sources of hydrogen production, 2020. Source: IEA (2021e).

There are, of course many different scenarios and considerable uncertainties. For example, in the analysis carried out by IRENA (2022), hydrogen could play a similar role to oil in transportation. Perhaps this view is too optimistic and the energy scenario will be far more mixed (see Figure 1.9)

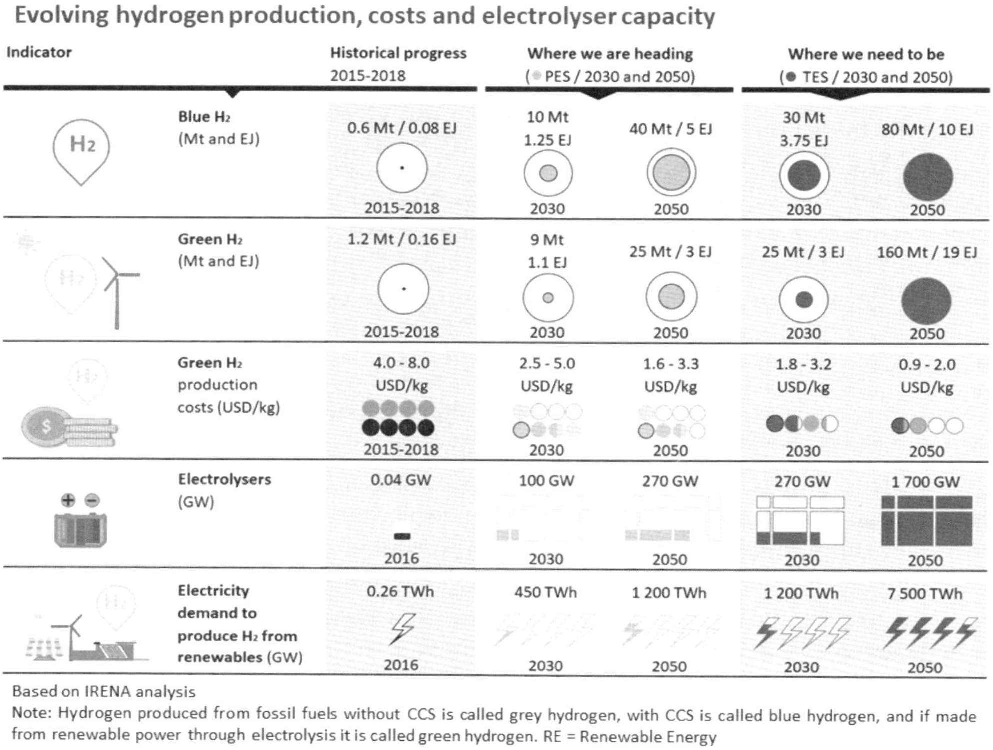

Figure 1.9 Hydrogen: A key part of future energy systems. Source: IRENA (2020).

1.10. EMERGING MARKETS FOR BIOFUELS (COMPETITION VERSUS COMPLEMENTARITY)

There are a large and diversified potential markets for biofuels beyond road transport. Some alternatives will be competing directly as fuel but in most cases, this will open up new opportunities, and be complementary to biofuels. This section will briefly consider cases where Brazil will have a competitive advantage: air transport, maritime transport, and fine chemicals (see following chapters).

1.10.1. BIOFUELS FOR AVIATION

About a decade or so ago, the academic community and a few of the major airlines began to show strong interest in "sustainable aviation fuels (SAF)" partly because the monopolistic structure of kerosene, rather than for environmental reasons. It was a wake up to a very tricky situation. In 2012 under the auspices of the IEA Task 40, a report was prepared (see ROSILLO-CALLE et al 2012) to investigate the potential use of biofuels in aviation. The main conclusions of the report ranged from an overall demand of 9EJ/yr. in 2010 to 16 EJ/yr. by 2050 (between 375 to 575 Mt). Assuming a 10% the potential use of biojet fuel, the demand was estimated at between 36.8 and 57.5 Mt (1.6 to 2.5EJ).

It turned out that the report largely underestimated these projections. Considerable amount of water has gone under bridge since then! Currently all major airlines are scrambling to find SAF, driven primarily by environmental considerations, but constrained by costs. Both IATA (2021) and ICAO (2022) have introduced specific policies to reduce emissions. In a few years, the vision and attitudes of air travel have shifted significantly in favour of achieving carbon neutrality. For example, IATA (2021) Annual Report indicates that SAF has the capability to reduce emissions by 80% in certain conditions. Despite high costs, SAF is increasing dramatically e.g., in 2021 production and use was estimated between 100 to 150 million litres (m/l), a 50% increase over the previous year. By 2025 about 5 B/l of SAF could be available, capable to meet 2% of global demand, and the 2030 projections are to cover at least 5% of global demand (IATA, 2021).

This trend is accelerating, with strong support from both IATA and ICAO. For example, ICAO in its 41st Assembly in October 2022, adopted a Long-Term Aspirational Goal (TAG) to achieve net zero CO_2 emissions by 2050 (ICAO, 2022a).

The interest on SAF has increased dramatically. Undoubtedly, this is a major potential market for biofuels and consequently requires clear and supporting policies on R&DD. Brazil was the first country in the world to have a policy on the use of biofuels in aviation (see BOEING/EMBRAER/FAPESP, 2013). Various studies have demonstrated the validly of biomass-based alternatives, particularly in Brazil e.g., see Walter et al (2021). There is a long way to go since aviation requires a very high-quality fuel. (See also Chapter 5).

1.10.2. BIOFUELS AND MARITIME TRANSPORT

This is perhaps one of the most promising emerging opportunity for biofuels in transport for various reasons: i) marine transport lags behind other sectors when it comes to pollution, ii) unlike aviation, the quality of the fuel is far lower and less stringent, iii) various studies have demonstrated it is possible to use agroforestry residues at competitive price e.g., Walter et al (2021) have carried out a study covering Brazil, Europe, South Africa and USA which demonstrates its viability. And in the case of Brazil, the study identified a potential of 196.5 PJ with costs ranging from $1.4 to $0.80GJ (WALTER et al, 2021). iv) this sector is under enormous pressure to cut pollution and catch up with other transportation sectors, with regards to the environment.

This is a growing sector, fundamental to world trade. It is estimated that four-fifths of trade (8 billion tons, out 9.84 billion) is transported by sea (UNCTAD, 2015). Other studies (CORTEZ et al, 2021; CARVALHO et al, 2021) have also shown the potential of using cellulose-based biofuels in maritime transport in Brazil (see Chapter 5).

1.10.3. BIOFUELS AND FINE CHEMISTRY (ETHANOL CHEMISTRY)

As most industrial sectors, the chemistry industry is moving away from fossil fuel-based chemical products and the focus is on renewable sources as much as possible. There are already many chemicals obtained from biomass e.g., fatty acids, lactic acid, pyrolytic biooil, specialty chemicals, butanol, bioplastics, glycerine etc. (e.g., see IEA, 2020).

Brazil used to have a small but flourishing ethanol-based fine chemical industry in the 1970s and 1980s (e.g., see ROSILLO-CALLE, 1986). This industry witnessed a gradual decline due to a number of factors, including i) direct competition from USA and Japan who had the market and know how, ii) lack of know-how in Brazil for this specialty sector; and iii) the growing petrochemical industry in Brazil.

Fine chemicals are a great emerging opportunity for the ethanol sector in the country because the feedstock can be produced at competitive cost and be available in sufficient quantities (see Chapter 5).

1.11. FOOD AND FUEL AND LAND USE

As stated previously, biofuels versus land use have been the subject of heating debate for decades. The "food versus fuel" debate needs to consider the intertwined nature of biofuels, land use, food production, food waste, animal feed production, diets, social habits, food distribution systems, policy, etc, as already stated. In fact, this controversy is being discredited, as explained in chapter 6. A more fundamental problem is unsustainable food production and lack of clear policy and debate on how to prioritize food systems.

It is almost unbelievable that agriculture is so low down in most government political priorities. Industrialization has received much greater investment. A study by the Global Alliance for the Future of Food (https://futureoffood.org) has put it very plainly, we need a radical change of how we produce food to save the planet.

Not only we produce food unsustainably, but we also waste a huge amount of it! About 17% of total food production is wasted (more than 900 Mt/year). And in some countries where refrigeration in transport is not available, as much as 75% of vegetables are wasted. A report by the Natural Research Defence Council says Americans throw away food worth $165 billion annually. It is an unbelievable waste so that the UN has started a strong campaign to reduce so much waste (see www.un.org/en/observances-end-food-waste).

In addition, according to UFOP (2022) most of grain production is used an animal feed rather than for direct human consumption, as illustrated in Figure 1.10. As can be appreciated, in 2021/22 harvest grain production was 2 333 Mt, of which 1 045 Mt were for animal feed (45%), food 749Mt (33%), and biofuels 168Mt or a mere 7%. The following figure (Figure 1.11) shows that about 81.6 Mha, or 7% of total cropland is dedicated to biofuels production. This does not consider the by-products obtained in biofuel production (e.g., animal feed, energy (from bagasse), yeast etc.).

In Brazil FvF has never been a real issue given the large land area (850 Mha) and abundance of uncultivated and underutilized land.

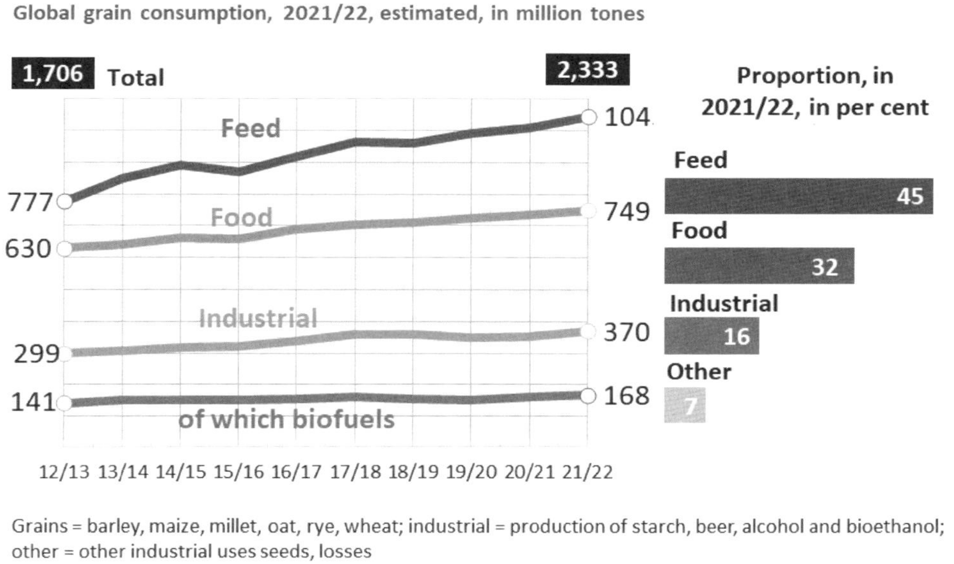

Figure 1.10 Grain is mainly used for feed production. Source: UFOP (2022).

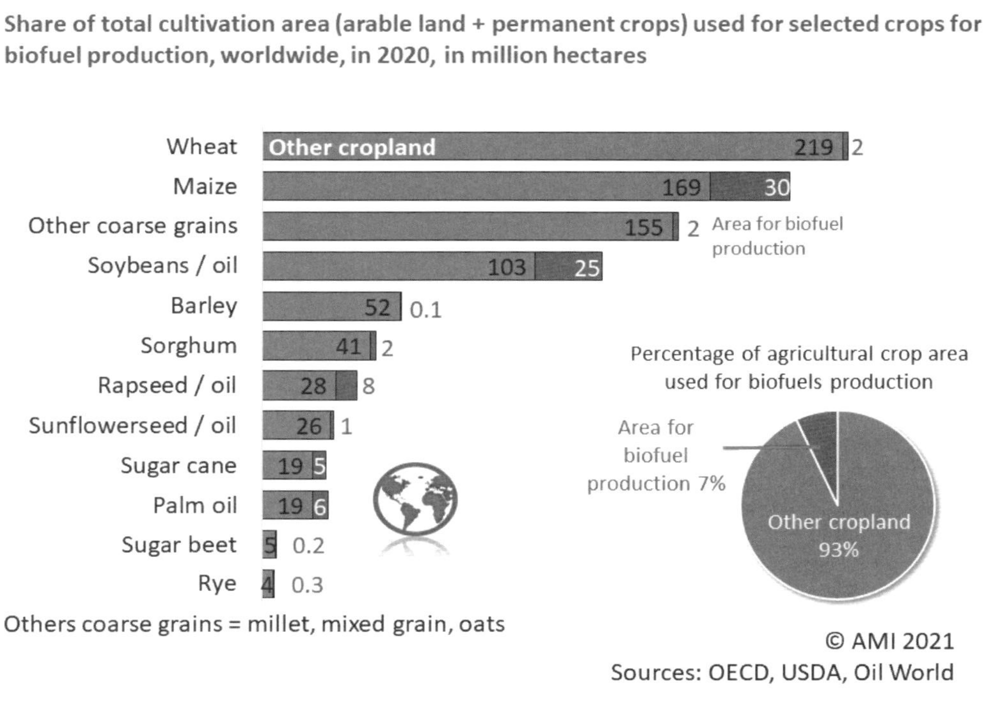

Figure 1.11 Biofuels take up little space. Source: UFOP (2022).

1.12. ENVIRONMENTAL ISSUES

Environmental sustainability and climate change will be key drivers of any future energy scenario. This requires international agreement or some consensus. An overly complex issue which we cannot do justice here given the complicity of issues involved (political, economic, social, technical etc). The implementation of biofuels, either nationally or in a global scale, thus depends on how such issues move forwards. Some countries will have greater autonomy than others. There is also the issue of the great unbalance among countries as to the amount of CO_2 emitted, as illustrated in Figure 1.12. While some rich countries as USA emits 15 tCO_2/person, in Brazil is a mere 1.9. This puts a lot of strain in poorer countries (see MME, 2021).

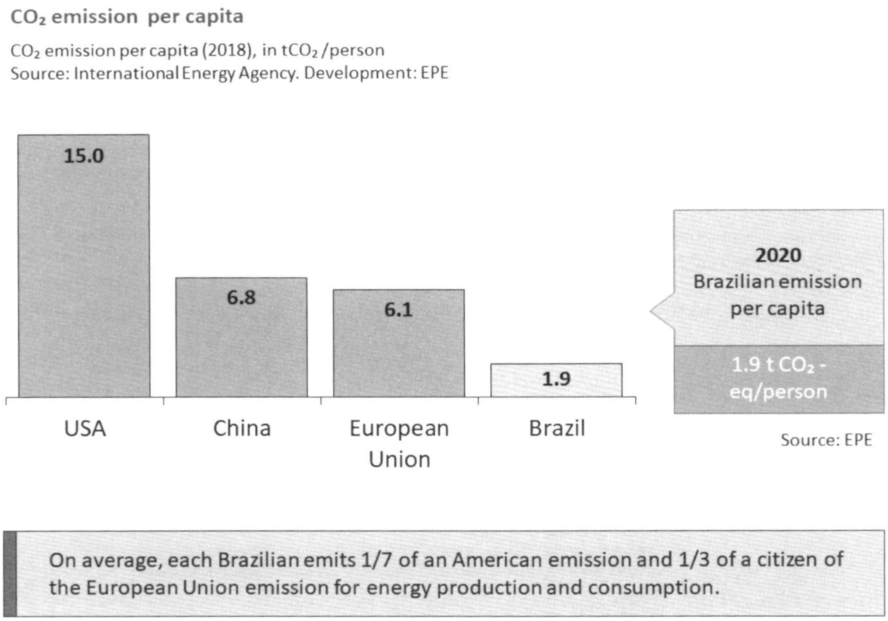

Figure 1.12 Brazilians per capita CO_2 emissions. Source: MME (2021).

FINAL REMARKS

This chapter has presented a brief summary of global energy scenarios, and potential implications. A shifting energy paradigm with huge potential impacts and ramifications, and the potential role of biofuels within this emerging reality. This overview will help the reader to better understand present energy trends and the implications for Brazil.

No country is an island and hence what happens around the world will impact directly or indirectly in an individual country and its policies. Even in Brazil with its unique position and historical experience and natural advantages, biofuels will not be immune to developments outside its borders.

Environment and climate change will be key drivers in any future anergy scenario as already emphasized. This requires international agreement or some kind of consensus. But environmentalism is clouded with conflicting and contradictory vested interests. Thus, the future development of biofuels will be partly shaped by external factors. And the question is *will Brazil use these advantages to promote biofuels in the future? Have biofuels a future in Brazil?* (see also Chapter 6).

The main challenge for Brazil seems political rather than technological, social, or economic, since the country has the technological know-how, natural and human resources, and research capability to deal successfully with any possible energy future scenarios.

This is what the following chapters try to answer.

REFERENCES

BOEING/EMBRAER/FAPESP. (2013). *Flightpath to Aviation Biofuels in Brazil*: Action Plan. BOEING, EMBRAER and FAPESP Project, June 2013.

BP. (2022). *BP Energy Outlook 2022*. Available Available at /.

CARVALHO, F.; PORTUGAL-PEREIRA, J.; JUNGINGER, M.; SZKLO, A. (2021). Biofuels for Maritime Transportation: A Spatial, Techno-Economic, and Logistic Analysis in Brazil, Europe, South Africa, and the USA, *Energies 2021*.

CIT. (2020). *Global Production of Advanced Liquid Biofuels* – Status Update, CIT – Norwegian Environmental Agency, by P Bokinhe and I Mystrom.

CORTEZ, L. A.; FRANCO, T. T.; VALENCA, G.; ROSILLO-CALLE, F. (2021). Perspective Use of Fast Pyrolysis Bio-Oil (FPBO) in Maritime Transport: The Case of Brazil, *Energies 2021*, 14, 479.

IATA – International Air Transport Association. (2021). International Air Transport Association. *Annual Report*. Available at www.iata.org.

ICAO. (2021). *Long Term Aspirational Goal* (LTAG), adopted during the 41st Assembly (October 2022).

ICAO. (2022). *Environment-Protection*; UN International Civil Aviation Organization. Available at https://icao.int/environmetal-protecion/.

IEA. (2017) *Technology Roadmap* – Delivering Sustainable Bioenergy. Available at www.ieabioenergy.com/.

IEA. (2019). *The Future of Hydrogen* – Seizing today's Opportunities. Available at www.iea.org/li.nzroadmap/.

IEA. (2020). *Bio-based Chemicals* – A 2020 Update. Available at www.ieabioenergy.com/.

IEA. (2021a). *Renewables 2021* – Analysis and Forecast to 2026. Report Extract – Biofuels. Available at www.iea.org/.

IEA. (2021b). *Global Energy Review 2021* – Analysis. Available at www.iea.org/reports/global-energy-review-2021.

IEA. (2021c). *Net-Zero by 2050* – A Roadmap for the Global Energy Sector. Available at https://www.iea.org/reports/net-zero-by-2050.

IEA. (2021d). *World Energy Outlook 2021*, IEA, Paris. Available at https://www.iea.org/reports/worldenergy-outlook-2021.

IEA. (2021e). *Global Hydrogen Review 2021.* Available at www.iea.org/.

IEA. (2021f). *Global EV Outlook 2021* – Prospects for Electric Vehicles Development. Available at www.iea.org.

IRENA – International Renewables Agency. (2020). *Global Renewables Outlook, Energy Transition 2050*: 2020 Summary. Available at www.irena.org/.

MME – Ministério de Minas e Energia. (2021). *Relatório Síntese 2020.* Balanço Energético Nacional, MME, EPE, Rio de Janeiro. Available at www.epe.gov.br/pt/publicacoes.

OECD/FAO. (2019). *OECD-FAO Agricultural Outlook.* Available at https://oecd.org/publications/oecd-fao-agriculturaloutlook/.

REN21. (2022). *Key Messages for Decision Makers.* RE 2021 Global Status Report. Available at www.ren21.com.

RFA. (2022). *Ethanol Outlook 2022.* Renewable Fuels Association. Available at https://ethanolrfa.org.

ROSILLO-CALLE, F.; JOHNSON, F. (2010). *The Food versus Fuel Debate*: An Informed Introduction to Biofuels, Zed Books, London.

ROSILLO-CALLE, F. (2018). Is really a food versus fuel dilemma? In *Sugarcane Bioenergy for Sustainable Development*, Routledge/Earthscan, Eds. L B Cortez, M RL V Leal, and L A Horta Nogueira, pp. 35-45.

ROSILLO-CALLE, F.; THRÄN, D.; SEIFFERT, M.; TEELUCKSINGH, S. (2012). *The Potential Role of Biofuels in Commercial Air Transport* (BIOJETFUEL), IEA Bioenergy Task 40. Available at www.bioenergytrade.org/.

ROSILLO-CALLE, F. (1986). *The Brazilian Ethanolchemistry Industry*: Biomass, Vol.11 (1) 19-38.

ROTHMAN, H.; GREENSHIELDS, R.; ROSILLO-CALLE, F. (1983) *The Alcohol Economy*: Fuel Ethanol the Brazilian Experience, Frances Pinter; and Kentucky University Press, Lexington, USA.

UFOP. (2022). *Annual Report 2022.* Available at www.ufop.de/reports.

UNCTAD. (2015). *Review of Maritime Transport 2015.* UN, Geneva. Available at www.unctad.org/systems/files/official-document/rmt2015_en.

WALTER, A.; SEABRA, J.; ROCHA, J.; GUARENGHI, M.; VIEIRA, N.; DAMAME, D. (2021). *Spatially Explicit Assessment of the Feasibility of Sustainable Aviation Fuels Production in Brazil*: Results of Three Case Studies, Energies, 14, 4972.

CHAPTER 2
HISTORIC BACKGROUND OF BIOFUELS IN BRAZIL

2.1. INTRODUCTION

When the Portuguese reached the Brazilian coast, probably the only source of energy they found was biomass. The country was named after a tree, the Brazil wood ("pau-brasil"), used to die clothes.

Sugarcane was officially introduced by Martim Afonso de Souza in 1532 and planted in São Paulo state coast (São Vicente) and other states in the Northeast region. Sugarcane was probably brought from Madeira Islands or Cabo Verde. Few years later, the production of sugar was successfully initiated by Duarte Coelho Pereira, in Pernambuco state.

About one or two centuries after introduction, sugarcane was mainly cultivated and developed in the Northeast region. Until 1580, the Dutch had direct involvement on sugar business produced in Brazil. They helped to finance the business as they also had an important role in the refining and commercialization of sugar in Europe.

Sugar was produced using the gur[1] or jaggery[2] method brought from India. At that time also "cachaça", the liquor from sugarcane, started to be produced. Cachaça, most likely, originated from the utilization of "cachaza[3]", the filter cake, as it is known today. Then, the cachaza was discarded by the engenho´s owner and then used by the first settlers. This filter cake is easily fermented and then distilled using an "alambique",[4]

---

1 https://en.wikipedia.org/wiki/Gur

2 https://en.wikipedia.org/wiki/Jaggery

3 https://publons.com/publon/3186206/

4 https://pt.wikipedia.org/wiki/Alambique

a still pot brought from Arabian countries to the Americas by the Portuguese and Spaniards.

The gur/jaggery sugar was produced first by extracting the juice (garapa[5]) and then boiling it up to crystallization point when "rapadura"[6] (a kind of gur/jaggery) was made. The boiling was made using metal open pans. This process was improved in a sequence of open pots in which the sugar juice was concentrated until the crystallization was completed (figure below).

Figure 2.1 The concentration of sugarcane juice prior to crystallization. Source: Leandro Vilar.[7]

After the juice was concentrated and ready to crystalize, the massecuite was poured on conic molds made of clay or wood, with a hole in the middle. After the crystallization was completed, the liquid molasses could be drained from the bottom and the sugar loaf could be unmolded. A curious observation is that the Portuguese noted the similar shape between the sugar loaf and the mountain located right in the entrance of Rio Bay naming the mountain after that (Pão de açúcar = Sugarloaf).

(a) (b)

Figure 2.2 (a) sugar loaf obtained from sugar crystallization; (b) The Sugarloaf Mountain in Rio.

5 https://pt.wikipedia.org/wiki/Garapa

6 https://pt.wikipedia.org/wiki/Rapadura#:~:text=Rapadura%20%C3%A9%20um%20doce%20de,pequenas%20quantidades%20para%20uso%20individual.

7 https://seguindopassoshistoria.blogspot.com/2013/12/o-engenho-e-o-fabrico-do-acucar-no.html?m=0

With the increasing production of sugarcane in the Caribbean in the 16 and 17th centuries, and the development of beet sugar in Europe, the Brazilian sugar production declined.

2.2. THE ETHANOL FUEL UTILIZATION IN BRAZIL

2.2.1. THE EARLY STAGES OF FUEL ETHANOL IN BRAZIL

During the second Brazilian Empire, D Pedro II created the Agronomic Institute of Campinas (IAC) in 1887 after his visit to USA (1st century of independence). D Pedro II was impressed with the agronomic research centres in the USA and wanted to do the same in Brazil. In the same year Franz Wilheim Dafert, the first IAC Director, presented 42 sugarcane varieties. The IAC research on sugarcane was initiated in 1892 which evolved, years later, into a more complete sugarcane variety program (SZMRECSÁNYI, 1979).

In early 20th century started a modernization process in the sugar industry with the objective to increase sugarcane yield and transform the old "engenhos" into modern mills. In the modernization process, "central mills" were created to gather sugarcane from the vicinities and centralize the processing. This fact created the preconditions to a more consistent development of the sugar sector in São Paulo state rather than in the Northeast states.

In 1903, the Brazilian President Rodrigues Alves inaugurated the *Exposição Internacional de Aparelhos a Álcool* in Rio de Janeiro, and the *Primeiro Congresso Nacional de Aplicação Industrial do Álcool.* In 1908, Henry Ford built the Ford T Model in the USA using ethanol and gasoline as its fuels.

Another parallel event that gave great impulse to the sugar industry, particularly in São Paulo state, was the Italian immigration. The Italians brought knowledge on how to make equipment and forged a strong industrial base with companies such as ZANINI and DEDINI.

During the 1920s several tests were carried out on the use of ethanol[8] as automotive fuel in Brazil. In 1925 a car successfully completed the journey between São Paulo to Rio using 100% ethanol as fuel (LEAL, 2022).

The Brazilian motor vehicle fleet developed rapidly reaching 220,000 in 1929, and consequently the consumption of imported gasoline.[9] With the urgent need to help

8 ethanol was first named "álcool", then "ethanol", and later "bioethanol". The reason for this change can be attributed to marketing since the molecule (C_2H_5OH) is the same.

9 The first oil refinery in Brazil, Landulpho Alves Refinery in Rio de Janeiro, was installed only in 1950 motivated by oil discovery in Bahia.

sugar producers together with the need to substitute, at least partially, imported gasoline, President Vargas signed the *Decree-Law n. 19.717*[10] in 1931. The main objective was to make compulsory= sugarcane ethanol to be added to the Brazilian gasoline. The blend was limited to 5% and varied according to the availability of ethanol.

During the WW II, when imported gasoline was lacking in Brazil, ethanol blends reached up to 50%. In 1942, President Vargas signed the *Decree-Law n. 4.722*[11] declaring the alcohol industry of national interest, establishing minimum prices for the product.

President Vargas also created in 1932 the Estação Experimental de Combustíveis e Minérios which became the Instituto Nacional de Tecnologia – INT, where Eduardo Sabino de Oliveira and Lauro de Barros Siciliano provided continuity to the studies initiated at the Escola Politécnica (Poli) at the University of São Paulo (OLIVEIRA, 1937; INT, 1979).

Also, in 1933, President Vargas created the Instituto do Açúcar e do Álcool (IAA) motivated by the crisis faced by the sugar sector. The main idea was for the IAA to act to establish *quotas* and arbitrate whenever necessary, both in the sugar and alcohol markets.

After the WW II, petroleum gained importance and in 1953 PETROBRAS was set up by *Law n. 2.004* during President Vargas second mandate. PETROBRAS was created as a state company to explore, refine oil, and with the mission to make Brazil oil self-sufficient. Later in the decade, President Juscelino Kubitschek gave great impulse to the automobile industry.

It is estimated that the average use of bioethanol between 1945-75 was around 7% (BNDES and CGEE, 2008), as seen in Figure 2.3.

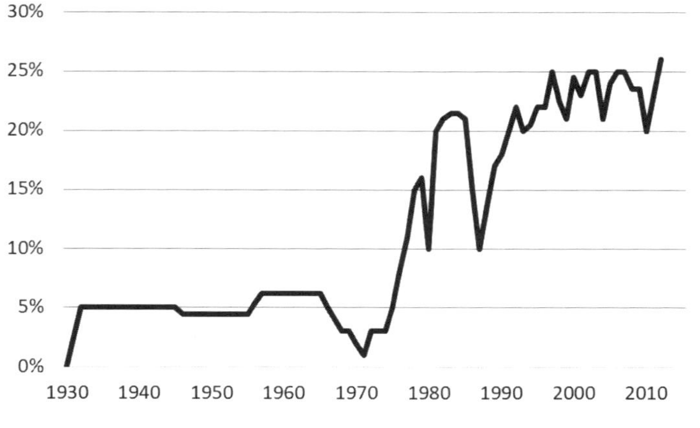

Figure 2.3 Ethanol content in Brazilian gasoline between 1930 to 2010. Source: MME (2008) in BNDES and CGEE (2008).

10 https://www2.camara.leg.br/legin/fed/decret/1930-1939/decreto-19717-20-fevereiro-1931-518991-norma-pe.html

11 https://www2.camara.leg.br/legin/fed/declei/1940-1949/decreto-lei-4722-22-setembro-1942--414753-publicacaooriginal-1-pe.html

The sugar sector was highly competitive during the 1960s since there was the Cuban revolution and the target to achieve 10 million tons of sugar by 1970 in Cuba. The result of the Cuban policy was a considerable increase in world sugar production and a corresponding decline in the price of sugar, predisposing the sector to another crisis.

At the end of 1960s the private sector created the COPERSUCAR Technology Centre (CTC) in Piracicaba. COPERSUCAR is a cooperative, which at that time, had more than 100 members. The CTC objective was to work on two key areas: an independent program to develop new sugarcane varieties, and a group of highly skilled agronomic engineers to assist the cooperative members and associates.

Also, in 1971 the Brazilian Federal Government created the National Program for Sugarcane Genetic Improvement (PLANALSUCAR). At the beginning of 1970s, the sector already had three major sugarcane improvement programs: the IAC, the CTC, and PLANALSUCAR.

2.2.2. THE 1973 OIL CRISIS AND THE CREATION OF PROALCOOL

The year of 1973 witnessed the first oil crisis with significant increase in oil prices, from US$ 1.9/barrel to US$ 11.2/barrel (see figure below). This fact deeply affected the Brazilian economy and politics. At that moment, oil imports represented nearly 80% of total oil consumption corresponding to nearly 50% of total imports. The situation was so critical that something had to be done urgently to help the economy and prevent the country coming to a halt.

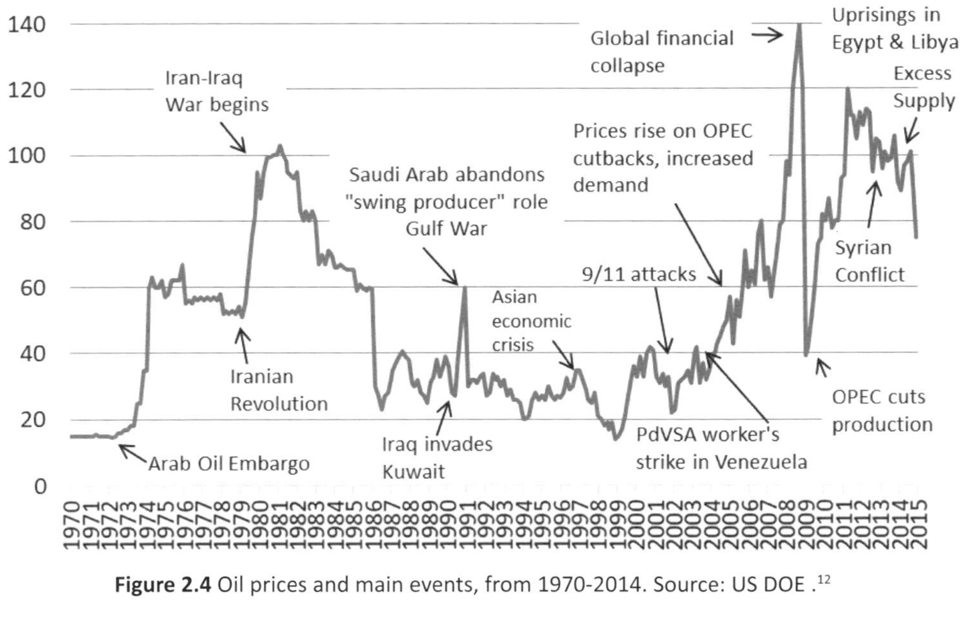

Figure 2.4 Oil prices and main events, from 1970-2014. Source: US DOE .[12]

12 https://theconversation.com/oil-price-shocks-have-a-long-history-but-todays-situation-may-be-the--most-complex-ever-178861

At that time, Brazil was experimenting a rapid economic growth, the so-called "economic miracle" under President Garrastazu Medici. General Ernesto Geisel, President of PETROBRAS, succeeded Medici and implemented a series of projects, promoting Brazilian energy independence. Among these projects were for example, the construction of several large hydroelectric plants, a nuclear energy program, and PROALCOOL, the National Alcohol Program.

PROALCOOL was created on 14 November 1975 by the *Decree n. 76.593* (MENEZES, 1980). It is difficult to indicate a single name as the PROALCOOL creator, and hence it may be fairer to mention few important names who helped to conceive and implement the program. Among them were: José Walter Bautista Vidal (engineer, physicist, and university professor), Lamartine Navarro Júnior (engineer, entrepreneur, and ULTRAGAS manager), Luiz Gonzaga Bertelli (consultant, lawyer, professor), Tobias J. Barreto de Menezes (university professor), Cícero Junqueira Franco (sugar mill owner), Expedito José de Sá Parente (engineer, chemist, creator of biodiesel), Ozires Silva (ex-president of EMBRAER, PETROBRAS, and minister of Infrastructure, Severo Fagundes Gomes (minister of Industry and Commerce), João Camilo Penna (also minister of Industry and Commerce), and Antonio Dias Leite Júnior (minister of Mines and Energy). See Cortez (2016).

Brazil was the first country to create a national biofuels program (PROALCOOL) and to produce large amounts of modern biofuels for fossil fuel substitution (ROTHMAN et al., 1983). At that moment the existing sugar mills were relatively small and the productivity, both agricultural and industrial, were low and something had to be done to improve it.

(a) (b)

Figure 2.5 (a) Minister Shigeaki Ueki fuelling ethanol in a car in Brazil; (b) President Ernesto Geisel (first on right) receiving entrepreneurs to discuss the PROALCOOL. Source: CORTEZ (2016).

In its origin, PROALCOOL tried to promote non-traditional energy crops such as cassava and sorghum without success. Probably the main reason why sugarcane succeed was the existence of a well-established sugarcane sector, particularly in São Paulo State. This sugarcane sector knew how to develop the new large-scale fuel ethanol industry, which was implemented in two steps:

The PROALCOOL first phase (1975-1979) had the objective to produce anhydrous ethanol to be blended with gasoline. According to ANP (2010), initially, 4.5% of ethanol was blended in the gasoline.[13] For this, annexed distilleries were installed in existing sugar mills. The production of ethanol increased from 600 million litres/year in 1975-76 to 3.4 billion litres/year in 1979-80. All investment was financed by the Brazilian government with subsided loans, both to expand sugarcane plantations and to install the annex distilleries. A great impulse of also given to increase the number of sugarcane varieties. At the beginning of the program North Argentina (NA) varieties were still predominant but they were gradually substituted by IAC SP and PLANTALSUCAR/RIDESA RB varieties as shown in the figure below.

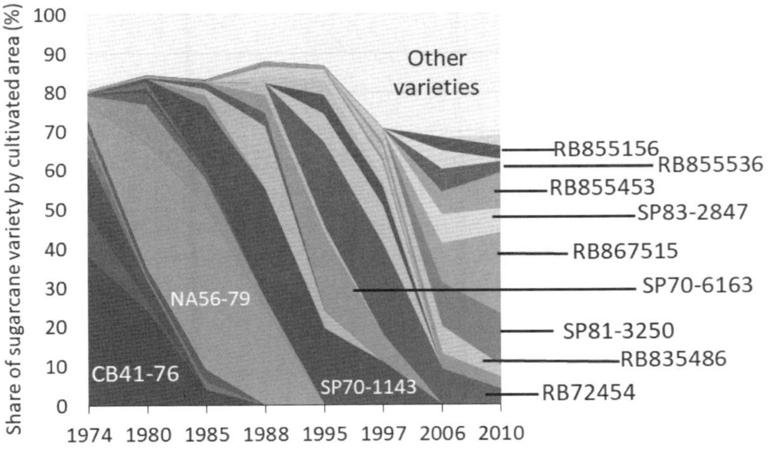

Figure 2.6 Evolution of commercial sugarcane varieties from 1974-2010 in Brazil. Source: CORTEZ (2016).

This decision to create and implement the PROALCOOL was not that simple. The idea was to improve the country´s national energy security. Therefore, directly, PROALCOOL didn´t have anything to do with greenhouse gases nor climatic change. The only objective was to alleviate the burden of imported oil and to support national sugarcane producers. The assistance to sugar producers was necessary due to serious problems such as low sugarcane productivity and an almost collapse of the sugar prices in the international market.

Initially, PROALCOOL intended to use not only sugarcane but also cassava and sorghum, as feedstocks. Sugarcane is a traditional large-scale crop in Brazil, primarily used for sugar production. Its main producing areas were the Southeast (São Paulo mainly) and the Northeast. However, in general terms, the sugarcane sector in São Paulo was at that point more dynamic and prepared to make good use of subsidies and other incentives than the Northeast, except for few mills in Alagoas.

13 https://www.researchgate.net/figure/Figura1-Evolucao-no-uso-dos-biocombustiveis-no-Brasil-Fonte-ANP-2010-Em-dezembro-de_fig13_320596282

On the other hand, cassava, although a traditional and indigenous Brazilian crop, despite good characteristics such as the easiness to store, did not take off. There were various problems e.g., cassava did not have a bagasse and hence the energy balance was negative. But probably the main reasons why cassava[14] failed as feedstock for ethanol in Brazil was the lack of agricultural experience with large scale plantations and inexistence of entrepreneurs interested to develop this crop, together with influence of the sugarcane lobby. There were also some serious problems in the industrial side such as difficulties with enzymes, infections, higher costs, etc. On the other hand, sugarcane was a better feedstock because it was a well-established, high productivity, and its bagasse could use as a by-product, (the fibrous by-product traditionally used to supply all factory energy needs in the mills). Cassava was a traditional crop, produced by many small holders, with low productivity while sugarcane growers, "*usineiros*", were a well-organized important group of rural entrepreneurs. The *usineiros* controlled several important cooperatives, associations and even had a well-organized R&D research centre(s) producing their own cane varieties besides exerting a considerable political and economic power.

Another crop that was not also successful was sweet sorghum, probably because of the same reasons as cassava.[15] A positive side effect of the PROALCOOL, or at least the intention, was to improve regional development, particularly with the introduction of other crops such as cassava and sorghum which turned out to have been unrealistic. It was believed, at that time, that with cassava and sorghum, the sugarcane monopoly could be avoided and regional development, particularly in the Northeast, could be achieved with the PROALCOOL.

The second phase of PROALCOOL (1979 to 1985) was a policy decision prompted by the second oil shock which forced the Brazilian government to speed up oil import substitution and put is faith into the new national fuel creating PROALCOOL II. This new phase was somehow more aggressive. According to ANP (2010) the percentage of ethanol blended in the gasoline went up to 15%.[16] Many completely new independent ethanol plants/distilleries were installed to produce only hydrous ethanol.[17] The new plants were conceived as autonomous distilleries and became the cornerstone because only ethanol was produced. The resulting hydrous ethanol was to be used in 100% ethanol vehicles (E100). This new phase was very successful, although there was an important subsidy associated, the so-called "ethanol account" which was compensated by the Brazilian government.

This situation could not continue after the democratic transition which took place in 1985, when the military government was replaced by a civilian one. This political transition set up the conditions for necessary changes in the economy which also impacted the recently created ethanol sector.

---

14 Thailand and Colombia verified to be more successful with cassava as feedstock for ethanol production.

15 There is considerable experience with the production of ethanol from cassava in Thailand. https://www.isaaa.org/kc/cropbiotechupdate/article/default.asp?ID=2348

16 https://www.researchgate.net/figure/Figura1-Evolucao-no-uso-dos-biocombustiveis-no-Brasil-Fonte-ANP-2010-Em-dezembro-de_fig13_320596282

17 According to the Brazilian National Petroleum Agency (ANP) the ethanol content of hydrous ethanol is between 95.1 and 96 ° GL or % by volume or between 92.5 and 93.8% by weight.

An important contribution was given by Ernesto Stumpf, from the Institute of Aeronautic Technology Institute) (ITA/CTA) who, with colleagues, investigated and developed the ethanol engine.

During that period, mainly 1975-1980, there were many engineering and non-engineering approaches to adapt the existing engines for ethanol blending. These blends were popularly known as "cocktails". Often, adaptations were made in garages without technical supervision.

However, the manufacturing industry was concerned with this issue, accepting that ethanol was becoming a reality in Brazil, and was to stay. Several adaptations were required to Otto cycle engines. A summary of these engine adaptations, performed by the automobile industry for ethanol blending in gasoline is given in the table below.

Table 2.1 Engine adaptations for ethanol blending in gasoline. Source: ANFAVEA[18]

Engine part Blending ratio	carburetor	Fuel injection	Fuel pump	Fuel pressure device	Fuel filter	Ignition system	Evaporative system	Fuel tank	Catalytic converter	Basic engine	Motor oil	Intake manifold	Exhaust system	Cold start system
<5%														
5 – 10%														
10-25%														
25-85%														

The Brazilian government wanted to stop paying royalties to manufactures (multinational automobile manufactures) who imported engines that were designed abroad. However, after the second oil crisis in 1979 the Brazilian government refused to accept their conditions and during an important meeting with ANFAVEA, made available a modified ethanol engine using E100, developed at the CTA. Mário Garnero, President of ANFAVEA, played an important role in this process too. The first 100% ethanol car was presented in 1979 by FIAT.[19]

Another important contribution was provided by an article published in Science under the title "Energy Balance for Ethyl Alcohol Production from Crops". The authors of this paper were José Gomes Silva et al (1978) This article stablished the basis for the negotiations involving sugarcane ethanol life cycle analysis (LCA).

Figure 2.7 below shows the main phases of the PROALCOOL including production volumes of ethanol and oil prices, 1972/1973 to 2014/2015.

18 Henry Joseph Jr. worked at Volkswagen and ANFAVEA playing an important role in these engine developments

19 https://www.automotivebusiness.com.br/pt/posts/noticias/fiat-comemora-40-anos-do-primeiro--carro-a-alcool/

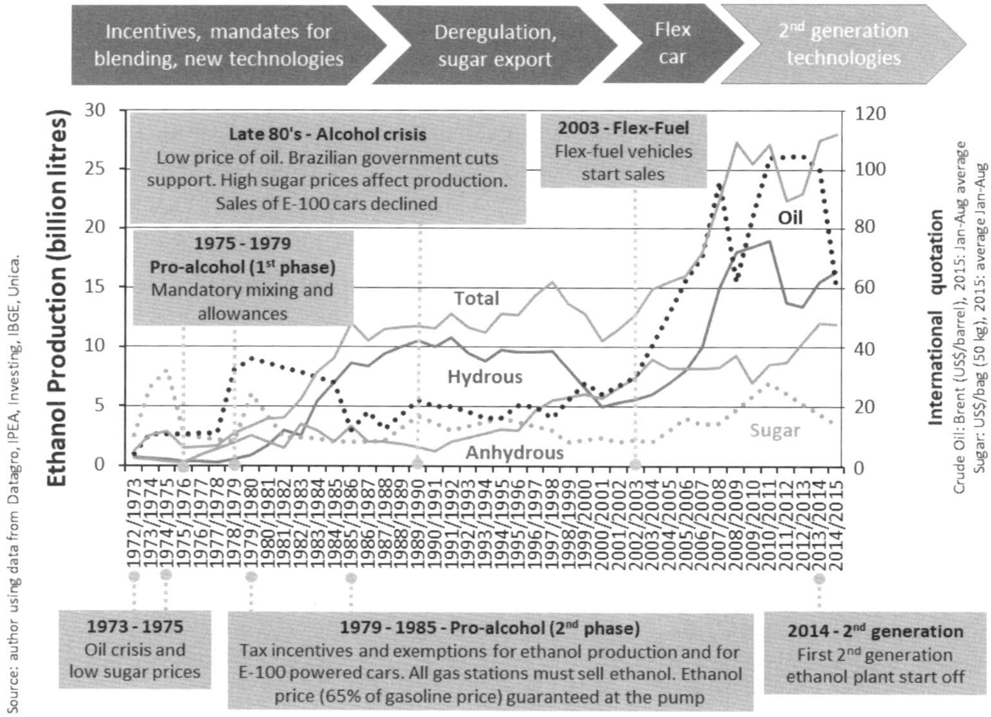

Figure 2.7 Main phases of PROALCOOL 1972-2015. Source: CORTEZ (2016).

2.2.3. THE NEW PERIOD (1985-2000): END OF SUBSIDIES TO ETHANOL AND INCREASING SUGAR EXPORTS

After 1985, the sugar-ethanol sector in Brazil went through an important transformation. The democratic government did not want to guarantee the alcohol account, so subsidies where lifted. The Institute of Sugar and Alcohol (IAA) responsible for regulating the market (quotas and other issues) was closed down as well as all incentives associated with ethanol production. Therefore, investment on new distilleries dropped fast.

The period 1985-2000 was also marked by low oil prices. This was a period of great financial difficulties for ethanol producers due to the removal of subsidies and lower gasoline prices. Even so, ethanol producers and PETROBRAS maintained the policy to blend ethanol in gasoline and continued to support E100 sales in gas/petrol stations all over the country. According to ANP (2010) the percentage of ethanol in gasoline reached 22% in 2015.[20]

The production of E100 light vehicles continued to raise despite the fact that no additional investments were made available to the sugarcane/ethanol production. This situation was aggravated to the new crisis in 1989 when there was a shortage of

20 https://www.researchgate.net/figure/Figura1-Evolucao-no-uso-dos-biocombustiveis-no-Brasil-Fonte-ANP-2010-Em-dezembro-de_fig13_320596282

anhydrous ethanol in the pump stations. The Brazilian government had to import fuel methanol to satisfy the demand. Methanol is considered more dangerous fuel to be handled than ethanol which also helped to irritate consumers. As a result, the 1989 crisis created a certain mistrust in the consumers and the E100 vehicles sales declined to almost zero.

In 1991 the IAA was closed down. This was not a minor change since years back the IAA was the organism responsible for quotas and regulation in the sugarcane sector. The closing down of IAA could be understood as an important move towards the deregulation in the sugar & ethanol sector in Brazil.

On the other hand, due to the ethanol crisis, more sucrose was used for sugar production which received a major boost. By the middle of 1990s Brazil was already responsible for almost half of all sugar traded in the world. The period 1990-2000 was marked by the sector deregulation and a significant increase on sugar exports; Brazil become the first sugar exporter in the world.

Surprisingly, despite the inconvenient of low oil prices in the 1990s, according to ANP (2010), ethanol blending with gasoline continued, varying from 20 to 25%.[21]

2.2.4. THE END OF SUGARCANE BURNING AND THE FLEX-FUEL ENGINE

The early years of 2000 was marked, from the very beginning, by an increasing concern on the environment. Sugarcane burning prior to harvesting was considered an important issue since it was happening exactly in the driest period of the year (May to October) provoking air pollution and increasing respiratory diseases. This was a serious concern as many sugar plantations were close to urban areas. José Goldemberg, then Secretary of Environment in São Paulo state, promulgated the *State Law n. 11.241/02* to abolish sugarcane burning. The period afterward was marked by a rapid technological transformation because, in practical terms, meant the end of manual cutting of sugarcane. This important event resulted in the full mechanization of sugarcane fields with the introduction of heavy machinery and logistics for harvesting operations. (See figure below).

21 https://www.researchgate.net/figure/Figura1-Evolucao-no-uso-dos-biocombustiveis-no-Brasil-Fonte-ANP-2010-Em-dezembro-de_fig13_320596282

Figure 2.8 Green sugarcane mechanized harvesting in Brazil. Source: Arthur Saraiva, site EMBRAPA.[22]

The beginning of XXI century was marked by the introduction of the flex-fuel vehicles in the domestic market, resulting in a rapid increase in ethanol demand. The sugar & ethanol sector had enjoyed its most prosperous years between 2002 and 2009 when it was growing at almost 10% annually.

Also, the sugar-alcohol sector paid much greater attention to improve energy efficiency in the mills. For example, the 2002 energy crisis forced more investments to improve cogeneration systems and created a third revenue from electricity sales for the mills, particularly in São Paulo state where several mills were connected to the national grid.

Since the implementation of the ethanol program, Brazil has created a sophisticated fuel ethanol regulation system comprising three phases:

- the first part was associated with sugar-ethanol distilleries. Most of sugar--ethanol distilleries in Brazil have the flexibility to produce either more sugar or ethanol as needed, depending on the circumstances e.g., the price of sugar or ethanol.
- Then, there is second part of the regulation system controlled by the Brazilian National Oil Agency (ANP). The ANP authorized the percentage of ethanol to be blended with gasoline as needed. So, when sugar prices are high, ANP authorizes less ethanol blending with gasoline and vice-versa.
- And thirdly the regulation system was a response by the consumer himself at the gas station. Depending on the price, the consumer will use hydrous ethanol (E100), or gasoline blended with anhydrous ethanol (today E27.5) depending on the price and his own car experience on mileage. On average, an ordinary car in Brazil will consume around 30% more of hydrous ethanol than if running on gasoline E27.5. Of course, the advent of flex-fuel engine and difficulties associated with increasing energy efficiency, somehow represent a limitation for the development of bioethanol in Brazil.

22 https://www.embrapa.br/busca-de-imagens/-/midia/4594001/colheita-de-cana-de-acucar-na-rmc

For further details on these three-level regulation system see Cortez et al. (2014) and on historical aspects of the PROALCOOL, Cortez (2016).

2.2.5. IMPORTANT TECHNICAL AND SCIENTIFICAL CONTRIBUTIONS OF THE SUGARCANE ETHANOL SECTOR

Several important contributions were made by the sugarcane ethanol sector since the implementation of PROALCOOL. The most important achievements in agriculture and industry are summarized in Table 2.2.

Table 2.2 Incremental technological advances by the sugarcane ethanol in Brazil.

Sector	Parameter	Previous status	Present status	Source
Agriculture	Sugarcane productivity	45 tons/ha in 1975	75-80 tons of cane stalks/ha today	sugarcane variety programs: IAC, PLANALSUCAR/RIDESA, CTC
	Sugarcane mechanization	only soil preparation was mechanized	Practically all field operations are mechanized	By the authors
	Green cane harvesting	All sugarcane was burned and hand cut	Practically all sugarcane is green harvested	
Industry	Juice extraction	93	97/98%	DEDINI
	Fermentation time	18 hours	6 hours	
	Fermentation yield	80%	92%	
	Wine ethanol content	6.5° GL	Up to 16° GL	
	Industrial yield	66 litres of ethanol/tc	87 litres of ethanol/tc	
	Steam consumption	620 kg of steam/tc	300 kg of steam/tc	
	Excess Bagasse	up to 8%	Up to 78%	
	Vinasse production	13 litres of vinasse/ litre of ethanol	5 to 0.8 litres of vinasse/litre of ethanol	
	Water use	262.5 litres/litre of ethanol	– 3.7 litres/litre of ethanol[23]	
	Agri-industrial productivity	3,000 litres/ha/yr	7,000 litres/ha/yr	
	Harvest season duration	6 months	8 to 11 months with sugarcane & corn	

23 This figure comes to be negative because part of water used in the process comes from the sugarcane itself.

Besides, there were several other contributions and improvements:

Agriculture:

- Increasing innovation in sugarcane planting
- Sugarcane diseases and pests' control
- Modern practices for sugarcane field management
- Implementation of highly competitive sugarcane logistics.

Industry:

- The size of sugar & ethanol mills grew up considerably ranging from 1 to 10 million tons of sugarcane/season (5 to 50,000 tons of cane/day) (crushed)
- Sugarcane dry cleaning, instead of water cleaning
- Increase in boilers efficiencies and cogeneration
- Maximizing electricity exports. For a 12,000 tons of cane/day plant, with 50% of bagasse/100% of trash excess, 84/112 MW of electricity.

Environment:

- GHG emissions reduction of sugarcane ethanol
- Important improvement of air quality in rural areas after the implementation of law stopping burning sugarcane
- Important improvement on air quality in Brazilian large cities

It is also important to mention the effort made by the productive sector to improve sustainability indicators of sugarcane ethanol in Brazil. An indication of this is given by the DEDINI concepts of optimization zeroing the amount of residues, odours, and water taken from water bodies. Also, great efforts have been made on mitigating GHG emissions comprising the whole production process.

Lastly, it is important to recognize that these efforts were compensated by declining costs (see the Ethanol Learning Curve below).

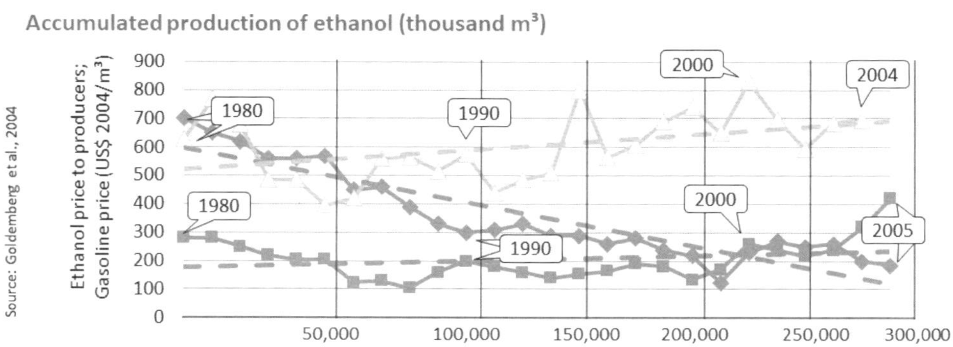

Figure 2.9 Sugarcane ethanol learning curve, price of ethanol according to the accumulated ethanol production volume. Obs: green-ethanol BR, yellow-gasoline BR, orange-gasoline Rotterdam. Source: Goldemberg et al (2004).

In addition, these gains were only possible thanks to an efficient partnership between the private sector, universities and research centres, together with government policies and incentives. Several of these contributions were made possible by institutions dedicated to science and technology in Brazil. To name a few: CENBIO, CTBE, EMBRAPA, INT, IPT and Universities e.g., UFRJ, UNESP, UNICAMP, USP.

2.2.6. THE PARIS AGREEMENT AND THE NEW ROLE OF ETHANOL IN BRAZIL

In 2015 the Paris Agreement, Brazil signed a commitment to reduce GHG emissions and consequently fuel ethanol regains its importance. In 2017 the federal government launches the RENOVABIO program with the objective of ensure the predictability in the fuel market, inducing energy efficiency gains and reducing GHG emissions on production, commercialization, and final use of fuels.

Overall, ethanol use in Brazil can be considered a success story with good prospects for continuity, particularly if higher engine efficiency could be achieved. Also, ethanol received considerable support from PETROBRAS. As stated by Nastari (2017) *"without ethanol PETROBRAS would not be able to commercialize its low-octane gasoline, and without PETROBRAS ethanol would not have been developed and distributed, so successfully"*. So, in general terms, there was until now a certain complementary effect of fossil-renewable biofuel in Brazil.

In general terms, it can be said that PETROBRAS benefited from ethanol in Brazil. Besides this issue presented by Nastari (2017), PETROBRAS could sell anhydrous ethanol at the same price of gasoline. In addition, as will be discussed later, PETROBRAS is still in a unique position to become a green-oil company, taking advantage of ethanol and other sustainable biofuels produced in Brazil

2.3. BIODIESEL AND THE PROBIODIESEL PROGRAM

The idea of producing biodiesel in Brazil is not new. At the end of 1970s, Expedito Parente from the University of Ceará, Brazil, worked on methods to obtain biodiesel and its utilization in engines for partial diesel substitution. The first patent was submitted to INPI in 1980 and accepted in 1983 (CORTEZ, 2016).

Melo and Fonseca (1981) refer to the difficulties associated with lack of an equivalent crop to sugarcane among the vegetable oils. While sugarcane overall yields are around 7,000 litres of ethanol/ha/yr, soybeans oil yield are circa 600 litres. Better yields can be found with palm oil with around 5,000 litres of oil/ha/yr in Indonesia or Malaysia.

Biodiesel production in Brazil was encouraged in the early years of this century and a program called PROBIODIESEL was set up by the *Law n. 11.097/2005* (TOKARSKI; ARANDA, 2019).

The initial objective of PROBIODIESEL was to produce biodiesel to be blended with conventional diesel oil. The main idea was to promote family production of non-traditional crops such as castor (Ricinus communis L.) and sunflower particularly in Northeast Brazil. However, despite all experimental trials and government incentives, castor oil could not be produced in large enough scale and high productivity, and hence the original ideal did not prosper.

The main initial questions regarding the biodiesel program were the feedstock, the dilemma faced by the right wing government interested to launch their own program without the so-called negative effect of PROALCOOL. On the industrial side, the vegetable oil was somehow an almost ready-to-use fuel with only need to eliminate the fat acids though transesterification process. Various projects attempted to overcome this problem e.g., at the State University of Campinas (UNICAMP), Schuchardt et al (1998) developed a transesterification process to allow the separation of fat acids in biodiesel production.

Geographically, the distribution of biodiesel plants coincides with soybeans production regions (mostly in Central-West and South) since it is the most important feedstock for biodiesel in Brazil, with c.70%, tallow 13%, and other feedstocks from vegetable or animal origin.

In 2021 Brazil produced 6.4 billion litres of biodiesel in 49 biodiesel plants, according to the Brazilian federal government,[24] but according to Torkarski and Aranda (2019) there were 51 biodiesel plants. The biodiesel program received important official financial support, particularly to install the industrial facilities which were relatively much less capital intensive when compared with ethanol distilleries from sugarcane. For many years the installed industrial capacity exceeded demand. As the years progressed, new mandates were introduced: B5 (2010), B6 and B7 (2014), B10 (2018), and B11 in 2021. Today the target is to reach B20 in 2030 with RENOVABIO (see Table 2.3).

Table 2.3 Evolution of biodiesel blends in Brazil. Source: UBRABIO based on 33.

Year	Biodiesel Blend (%)
2017	8
2018	9
2019	10
2020	11
2021	12
2022	14

24 https://www.gov.br/pt-br/noticias/energia-minerais-e-combustiveis/2021/07/brasil-avanca-no-setor--de-biocombustiveis

Year	Biodiesel Blend (%)
2023	16*
2024	18
2025	20 (51% of INDC target)
2026	22
2027	24
2028	26
2029	28
2030	30 (80% of INDC target)

*Obs: In 2023 the blend is actually still 10% of biodiesel.

Biodiesel does not have a good output/input energy ratio as ethanol. In Brazil, according to Macedo and Nogueira (2004) the ratio ranges from 2-3 for biodiesel and 8.3 for sugarcane ethanol. However, for palm oil and macaúba, Martins and Teixeira (1985) expect the ratio to be 5.63 and 4.20, respectively. These crops present a higher output/input energy ratio. However, the difficulties with both dendê palm (*Elaeis guineensis*) and macaúba (*Acrocomia aculeata*), is that they are not traditional large-scale crops in Brazil, therefore, there is not sufficient reliable agronomic experience for large scale cultivation. Figure 2.10 shows the evolution of plant capacity and consumption, from 2008 to 2019.

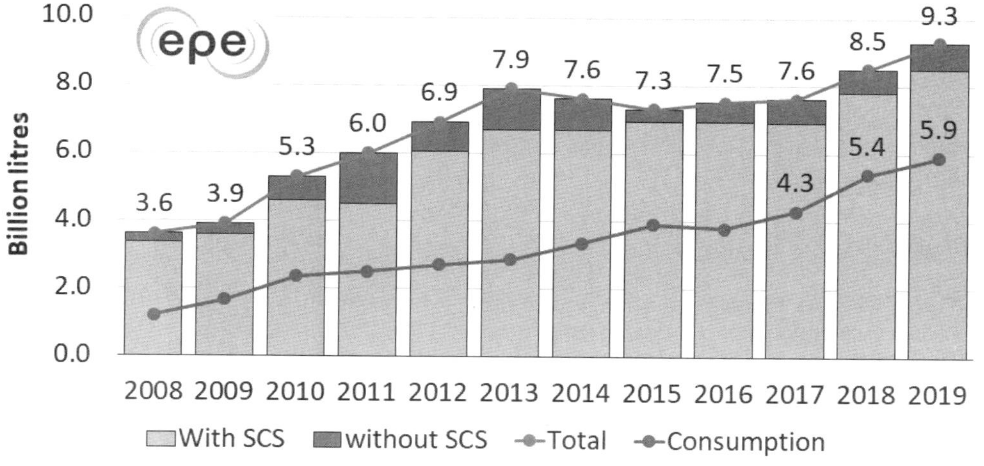

Source: EPE from (EPE, 2020a) and (ANP, 2020c).
Note: The Social Fuel Seal (SCS) is a distinction given to companies that produce biodiesel that use products from Family farms in their production chain. The objective is to guarantee income and stimulate the social inclusion of producing families. Biodiesel producers and SCS holders benefit from access to better financing conditions from financial institutions.

Figure 2.10 Authorized nominal capacity and biodiesel consumption in 2019. Source: EPE (2020).

Greater increments on blending are expected in the near future, as it will be discussed on the next chapter.

2.4. LESSONS OF BIOFUELS FROM THE PAST 50 YEARS IN BRAZIL

The cumulative contribution of ethanol and biodiesel to Brazil (Figure 2.11) can be considered very important. Its contribution is not only important from an energy and macroeconomic perspective, but also from a social and environment points of view. Although institutions such as the EPA from USA has recognized sugarcane ethanol produced in Brazil as an advanced fuel, its global contribution has not yet been fully recognized.

Besides the macroeconomic benefits derived from hard currency savings, a significant number of direct and indirect jobs were created by ethanol and biodiesel industries in Brazil. Up to 2000s, a large number of jobs were on sugarcane harvesting. However, after the non-burning law has been approved, the sector gradually started to use mechanized harvesting substituting manual labour. At the beginning there was a concern, particularly by the rural unions, because manual harvesting attracted a large number of workers from different regions of the country. However, as the sector expanded, requiring more workers, these workers were absorbed in other sugarcane and ethanol activities. Today, the biofuels industry is one of the most important contributors to economic prosperity in rural Brazil. For example, it has been estimated that there are nearly one million workers, direct and indirect. Just the sugarcane sector alone employs more than 600,000 workers (AZANHA, 2010).

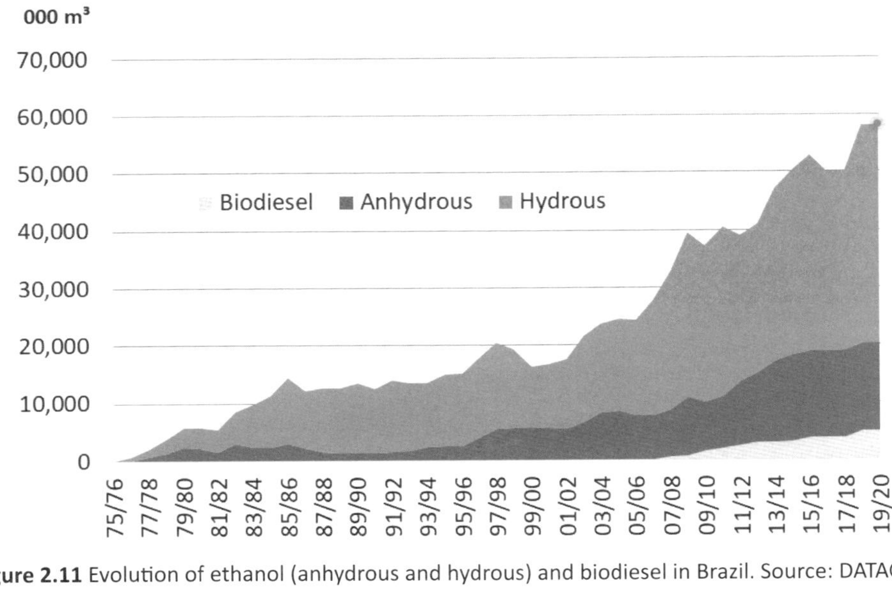

Figure 2.11 Evolution of ethanol (anhydrous and hydrous) and biodiesel in Brazil. Source: DATAGRO and ANP, from Leal (2022).

Although the use of modern biofuels started almost 100 years ago, it can be said that the most intensive use happened in the last 50 or so years (since 1975), particularly with ethanol from sugarcane. From that, several lessons can be derived:

- **The successful use of biofuels depends on three things: feedstock production, feedstock conversion process, and biofuels final use (engines).** So, the three must be efficient, not just one or two. Feedstock efficient production needs also to be competitive and environmentally sustainable. Feedstock conversion process needs to be energy efficient and sustainable with corresponding use of all by-products. And lastly, biofuels final use should also be efficient. Overall, it is an equation in which all the three components (feedstock, process, and engine) must be efficient and competitive.
- **Biofuel economics are heavily dependent on the feedstock**. For ethanol this represents around 70%, and for biodiesel c.80%. So, if the economics of feedstock is not well established, there is little that can be done.
- **Biofuel economics also depend on the valorisation/utilization of by-products**. This happens with sugarcane ethanol, in which sugar (for food) acts as the main coproduct, and ethanol for fuel and bagasse (for electricity) as second. The same phenomena occur with corn ethanol in which DDG,[25] the main coproduct from corn ethanol, is used as cattle feed. This point will be discussed in more detail below.
- **If biofuels are conceived to reduce GHG emissions, special attention should be given to its life cycle analysis**. For example, ethanol from sugarcane or corn ethanol, can be produced in different ways, hence achieving different life cycle results.
- **If biofuels are conceived to promote social benefits, its premisses should be analysed**. Ethanol or biodiesel, or any other biofuel, needs to be produced at competitive cost to be able to compete with fossil fuels. Markets are not willing to pay premium costs to any biofuel, unless forced by legislation. Of course, there is the problem posed by heavy subsidies given to fossil fuels, and in this case, oil. These subsidies to fossil fuels should be eliminated to increase the competitiveness of biofuels and other renewables to have a fairer playing field
- **Indirect land use change (iLUC) is a controversial concept** (to be further discussed on Chapter 4). It is not proven that a modern biofuel such as sugarcane ethanol produced any iLUC, at least in Brazil, **nor any negative impact of ethanol, nor biodiesel to food production**. On the contrary, sugarcane ethanol benefited sugar, and modernized other agricultural activities; soya biodiesel production benefited poultry or swine production in Brazil, as corn ethanol benefited cattle production in the USA.
- **All residues need to be recycled**

25 DDG-dry distillers' grains.

FINAL REMARKS

This chapter has highlighted how biofuels have evolved in Brazil through its various historical phases. A special attention was given to fuel ethanol. It has identified the most salient features such as policy changes, technological achievements, regulations, major actors in the development of biofuels, agriculture, social, and environmental issues, as well as lessons learned.

It has been a long road since environmental sustainability was taken seriously. In the old school of thought, economic development was at the core of policymaking in Brazil, and biofuels development reflects this reality.

This chapter provides a realistic overview of this historical process for a spectrum of readers.

REFERENCES

AZANHA, M. *More Jobs with Ethanol* (in Portuguese). Available at https://www.investe.sp.gov.br/noticia/mais-empregos-com-o-etanol/.

BNDES/CGEE. (2008). *Sugarcane-based Bioethanol for Sustainable Development.* First edition, Rio de Janeiro, 304 p. Available at https://web.bndes.gov.br/bib/jspui/bitstream/1408/6305/1/2008_Sugarcane-based%20bioethanol_energy%20for%20sustainable_P.pdf.

CORTEZ, L. A. B. (coord.) (2010). *Sugarcane Bioethanol R&D for productivity and sustainability*, Blucher, 992 p.

CORTEZ, L. A. B.(coord.) (2016). *Proálcool 40 Anos.* Universidades e Empresas: 40 Anos de Ciência e Tecnologia para o Etanol Brasileiro. Blucher, 233p.

CORTEZ, L. A. B.; SOUZA, G. M.; CRUZ, C. H. B.; MACIEL, R. (2014). An Assessment of Brazilian Government Initiatives and Policies for the Promotion of Biofuels Through Research, Commercialization, and Private Investment Support. In *Biofuels in Brazil*, Springer, p. 31-60. Available at https://link.springer.com/book/10.1007/978-3-319-05020-1.

CORTEZ, L. A. B.; LEAL, M. R. L. V.; NOGUEIRA, L. A. H. (2019). Sugarcane Bioenergy for Sustainable Development Expanding Production. In *Latin America and Africa*, Routledge, 2019, 444 p.

DA SILVA, J. G.; SERRA, G. E.; GONÇALVES, C.; GOLDEMBERG, J. (1978); Energy balance for ethyl alcohol production from crops, *Science*, 201, 4359: 903-906.

DE ANDRADE NETO, J. L. (2008). *Biocombustíveis*: a experiência brasileira e a visão do governo. PPT Presentation, Ministry of Mines and Energy, Brasília, DF, 27 August 2008. Available at https://slideplayer.com.br/slide/381291/.

EPE – Empresa de Planejamento Energético. (2020). *Analysis of Biofuels´ Current Outlook – Year 2019.* EPE/Ministry of Mines and Energy, Rio de Janeiro, July 2020, 80 p. Available at https://www.epe.gov.br/sites-pt/publicacoes-dados-abertos/

publicacoes/PublicacoesArquivos/publicacao-489/Analysis_Biofuels_Current_Outlook_2019.pdf .

GOLDEMBERG, J. et al. (2004). Ethanol learning curve – the Brazilian experience. *Biomass and Bioenergy*, v.26, n.3, p. 301-3004, March 2004.

LEAL, M. R. L. V. (2022). *Etanol*: Desenvolvimento Tecnológico, Inovação, Sustentabilidade, Zoom Meeting – AEITA, March 2022.

MACEDO, I. C.; NOGUEIRA, L. A. H. (2004). *Avaliação do Biodiesel no Brasil*, CGEE, Brasília, DF, 48 p.

MARTINS, H.; TEIXEIRA, L. C. (1985). Balanço energético da produção de óleos vegetais transesterificados, *III Congresso Brasileiro de Energia*, Rio de Janeiro.

MELO, F. H. de; FONSECA, E. G. da. (1981). *Proálcool, Energia e Transporte*. Estudos Econômicos FIPE/Pioneira, São Paulo, 163 p.

MME – Ministério de Minas e Energia. (2008). *Balanço Energético Nacional*, MME, EPE, Rio de Janeiro. Available at www.epe.gov.br/pt/publicacoes.

NASTARI, P. (2017). *Compreendendo o RENOVABIO* (Understanding the RENOVABIO). Available at https://www.fiesp.com.br/arquivo-download/?id=232631.

ROTHMAN, H.; GREENSHIELDS, R.; ROSILLO-CALLE, F. (1983). *The Alcohol Economy*: Fuel Ethanol the Brazilian Experience, Frances Pinter; and Kentucky University Press, Lexington, USA.

SCHUCHARDT, U.; SERCHELI, R.; VARGAS, R. M. (1998). Transesterification of vegetable oils: a review, Review Article. *J. Braz. Chem. Soc.* 9 (3), May 1998. Available at https://doi.org/10.1590/S0103-50531998000300002.

SZMRECSÁNYI, T. (1979). *O Planejamento da Agroindústria Canavieira do Brasil (1930-1975).* São Paulo, Hucitec, 540 p.

TOKARSKI, D.; ARANDA, D. (2019). Biocombustíveis: Histórico e Oportunidades (Biofuels: history and opportunities), UBRABIO, *Seminar Biocombustíveis: A nova realidade do Brasil.*

CHAPTER 3
THE PRESENT SITUATION OF BIOFUELS IN BRAZIL

3.1. INTRODUCTION

Typically, every time Brazil faces an energy crisis (oil increase or water/electricity shortages) ends up with a new emphasis to expand RE, in particular biofuels. This was the case during the energy crisis of 1920s, WWII, 1973, 1979, 2007, and now in 2022. Different reasons precipitated these crises, all of them affecting Brazil, both from the energy or the economic points of view. The difference now is that new increase of oil prices, due to the Russian war on Ukraine, coincides with an energy transition process that will take decades.

Energy wise, Brazil is currently a country with a relatively low per capita use. According to IEA (2021) the country is the world's 8th largest energy consumer but the 69th largest energy per capita.[1] In addition, traditionally, Brazil has made considerable progress in renewable energies, both for electricity production and transport. In the Brazilian energy matrix, the contribution of renewables was 48.1% in 2020. In electricity production, 83% is renewable[2] and in transportation, ethanol represents 43% of fuel in light vehicles, and biodiesel 10% of fuel in heavy vehicles. (MME, 2021). Brazil is, therefore, the greatest renewable energy economy in the world (see Table 3.1).

1 https://worldpopulationreview.com/country-rankings/energy-consumption-by-country

2 According to the Brazilian Government energy statistics (MME, 2021) 62.5% is hydro, 9.8% wind, 8.8% biomass, and 1.9% solar photovoltaics.

Table 3.1 Production of modern bioenergy in Brazil in 2021. Sources: as indicated in the table.

Type of biofuel	Feedstock	Production	Policy	Source
bioethanol	Sugarcane (90%) Corn (10%)	29.2 billion litres	E27.5 and E100	3
biodiesel	Soybeans (90%)	6.4 billion litres	B10	4
Bioelectricity	Sugarcane bagasse	9.1% of total electricity generated	-------	MME (2021)

The use of ethanol has intensified in the last two decades, particularly after the introduction of flex-fuel engines in 2003. Figure 3.1 shows ethanol (hydrous and anhydrous) for the 2003-2007 period. This gives an idea how fast this change has occurred. As the figure shows, the consumers initially accepted well the flex-fuel innovation, although currently expectations are much lower.

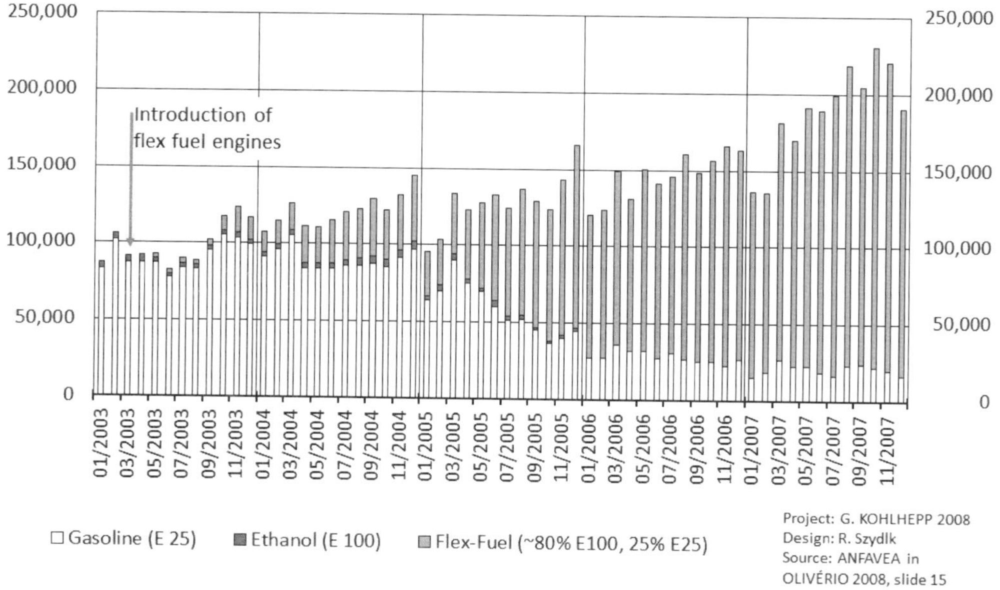

Figure 3.1 Evolution of gasoline (E25) and hydrous ethanol (E100) consumption in Brazil depending to the engine type. Source: ANFAVEA, cited by Olivério (2008).

Therefore, it can be stated that Brazil is the country among the 10 largest economies with the best profile in renewables, both for electricity production and liquid biofuels. No other economy, despite its technological development compared to advanced economies, has made so much progress in encouraging renewable energies, biofuels included, that Brazil.

3 https://www.gov.br/pt-br/noticias/agricultura-e-pecuaria/2021/10/producao-de-etanol-de-milho--avanca-no-pais-como-opcao-sustentavel-e-de-valor-agregado

4 https://www.noticiasagricolas.com.br/noticias/politica-economia/272575-producao-de-biodiesel-do--brasil-cresce-85-em-2020-diz-abiove.html#.YfA1fOrMI2w

So, the issue in energy transition in Brazil can be better understood by growing overall energy production while maintaining the relative importance of renewables. As a trend in energy transition for electricity, Brazil has excellent possibilities with solar, wind, and biomass. Regarding liquid biofuels there is still a room for much greater contribution of bioethanol and biodiesel, both for domestic market and when exports are considered. Also, if future markets such as biofuels for aviation and maritime are included, the prospects are also very good.

The opportunities are also very good concerning the production of new biomaterials, the so called "green" materials and fine chemicals. An example is the bio-based polyethylene wax developed by Braskem.[5] This new renewable material open possibilities in multiple markets, such as adhesives, cosmetics, coatings and masterbatches.

Of course, the realization of these opportunities will greatly depend on how the country successfully deploys the existing relative advantages, such as available fertile land and other natural resources, together with human capacity and capital.

3.2. CURRENT SITUATION OF ETHANOL IN BRAZIL

The Brazilian Ministry of Agriculture (MAPA) registered in 2020, 361 sugar & ethanol producing plants, with a capacity to process 745 million tons of sugarcane (EPE, 2022). The same source estimates the total sugarcane harvested area will grow from 8.3 Mha in 2021 to 9.3 Mha in 2031. Productivity is expected to increase by 1.4%/year, reaching 83 tc/ha in 2031, while the yield should be around 141 kg of Total Recoverable Sugars (TRS)ATR/tc.

3.2.1. AN OVERALL VIEW OF PRESENT STATUS OF ETHANOL

Ethanol is currently known as bioethanol, meaning of biological origin, as well as biodiesel. This renaming simply emphasizes its biological nature rather than any chemical change. In recent years, ethanol supply has not been affected as much as it was often the case in the past. The price of fuel ethanol (E100), however, is closely linked to price variations of gasoline (E27.5) (see Figure 3.2)

This, of course, discourages the consumer, particularly those located far from the production regions. Flex fuel engines do not represent, in practice, an economic advantage to the consumers. And so, it seems, would be the case for 2G ethanol in the near future unless the economics can be improved significantly. Corn ethanol, on the contrary, is expanding fast, benefiting consumers from Mato Grosso and Goiás states. Nevertheless, productivity will have to improve significantly to be able to compete with sugarcane ethanol producing regions.

This chapter assesses the present status of biofuels in Brazil with the objective of analysing its possible role in the energy transition which the world is going through.

5 https://www.braskem.com.br/imgreen/bio-based-en

The main argument is that sugarcane ethanol, together with biodiesel, represent a great contribution for GHG emissions mitigation, and consequently there aren't any reasons to adopt other strategies. To explain this, following is an analysis of each parameter of the production phase and biofuels GHG emissions, in relation to the commitments signed by Brazil in the Paris Agreement of 2015.

Figure 3.2 shows estimated ethanol production (hydrous and anhydrous) from 2021 through 2031. As can be appreciated, the increase is primarily hydrous ethanol. Figure 3.3 illustrates ethanol trade (domestic and international) from 2008 to 2019, according to EPE estimates.

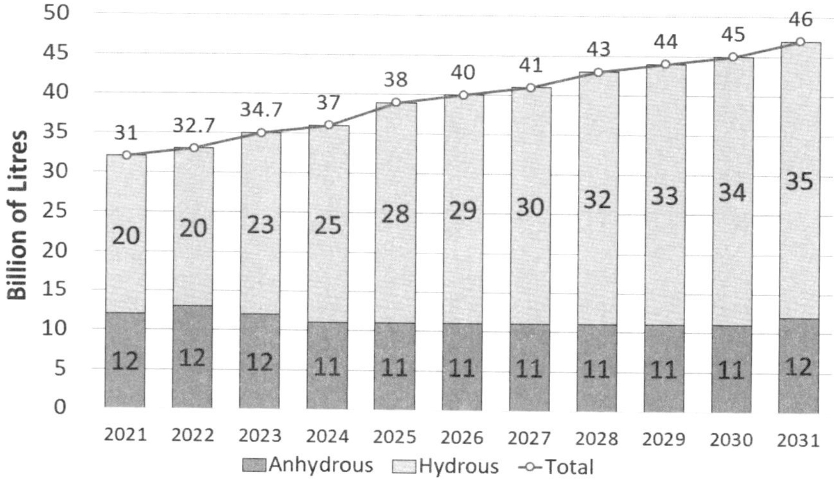

Figure 3.2 Present and future estimations of ethanol production and imports from Brazil. Source: EPE based on EPE (2017).

Although there is a concern of ethanol imports, historically it represents less than exports (figure below).

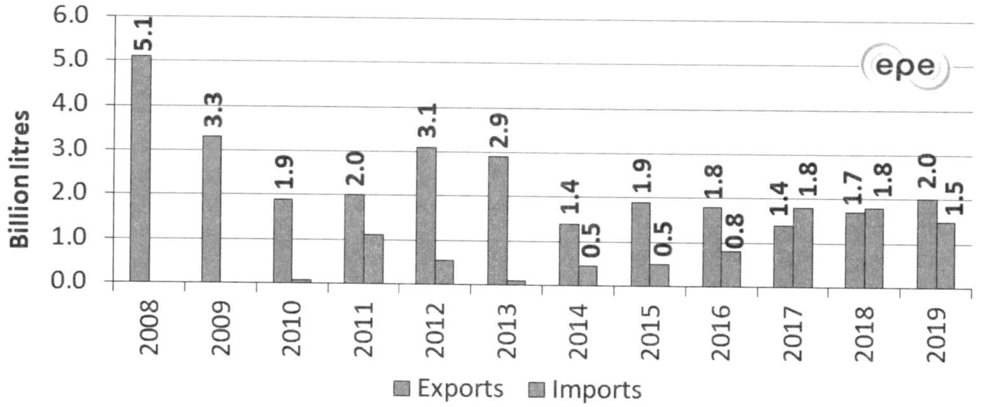

Figure 3.3 Brazilian exports and imports of ethanol from 2008 to 2019. Source: EPE (2020) from MME (2020).

3.2.2. THE PRESENT STATUS AND OPPORTUNITIES FOR SUGARCANE

As stated before, for a biofuel to be competitive, the feedstock needs to be produced in a competitive environmentally sustainable manner since, typically, it is responsible for 70-80% of the overall cost.

As stated in the previous chapter, Brazil has made a considerable effort to diversify sugarcane varieties to increase productivity, drought resistance, and new and more adapted varieties to increase the harvesting period.[6] It is important to state that present productivity, around 75 tons/ha/yr, is relatively small because of reductions associated with adverse climatic conditions aggravated by the introduction of harvesting mechanization which requires heavy machinery causing soil compaction. Figure 3.4 below shows the share of cane going to ethanol and to sugar production together with estimated cane productivity from 2021 through 2031.

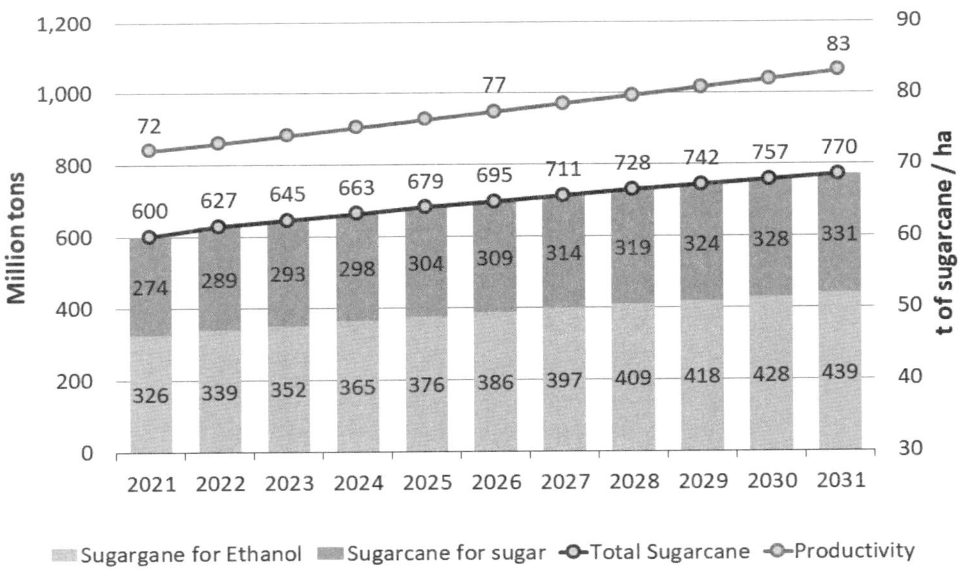

Figure 3.4 Productivity, harvested cane, and destination to ethanol and sugar. Source: EPE (2022).

Therefore, of the 9-10 Mha dedicated to sugarcane production, nearly half goes to sugar and the other half to ethanol. With nearly to 4.5 Mha equivalent dedicated to sugar, Brazil has become the leading exporting country in the world (Figure 3.5 below).

6 The first harvest typically occurs 18 months after the cane stalks are planted. Then, the following harvests occurs after 12 months. Typically, the harvesting yields decline making necessary to replant the cane after 5-6 harvests.

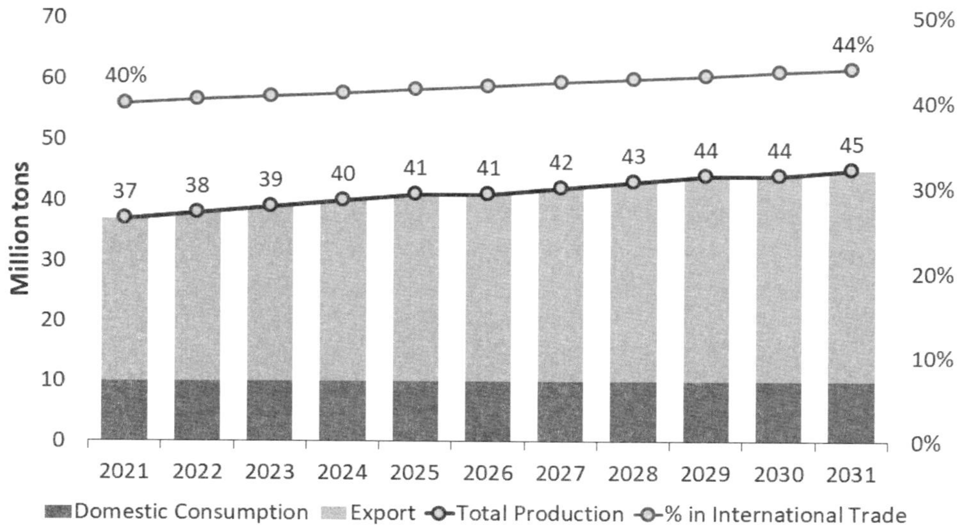

Figure 3.5 Projection of Brazilian production of sugar until 2031. Source: EPE (2020).

A major recent improvement with sugarcane as feedstock (for ethanol, electricity and 2G ethanol) was the introduction and development of energycane,[7] a cane variety that proportionally produces more fibre than the conventional cane. It has been tested by GranBio in Alagoas in its 2G ethanol production plant. Energycane could also be, potentially, an excellent source for electricity production. One negative side of energycane is that its high fibre content renders the extraction more difficult, requiring more energy and equipment maintenance. Hence more technical improvements are required in preparing the cane prior to extraction. As for sugar, energycane produces less sugar per weight but more sugar by area.

With the introduction of flex mills, operating with sugarcane and corn, will also require more studies on energy integration. Equipment companies such as DEDINI are aware of the problem and the opportunity it represents and is looking for solutions. Corn ethanol is expanding in Brazil, as it will be seen later in this book.

Figure below (3.6) shows the map with the main localities of sugarcane plantations in Brazil.

7 According to EPE (2022) energycane will represent around 100 thousand ha, corresponding to 1.1% total sugarcane area in 2031.

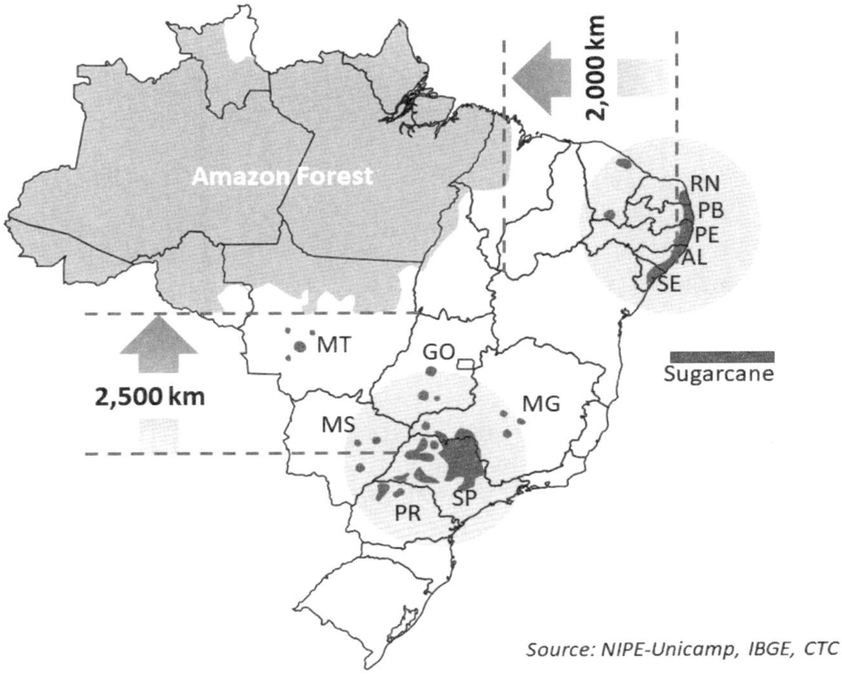

Figure 3.6 Map of sugarcane distribution in Brazil. Source: NIPE-Unicamp; IBGE; CTC (2017).

Sugarcane has several positive features e.g., more resilient to climate change, good soil coverage preventing erosion, high fibre content besides sugar, adaptation to different soil conditions, and very good productivity when compared to other crops.[8] On the other hand, sugarcane needs to be processed quickly after its harvest to prevent sugar losses, while its harvesting season typically covers 7-8 months in Southeast and Centre-South. The introduction of green cane mechanical harvesting brought also problems associated with diseases and soil compaction. The later, together with droughts, are probably the most important factors responsible for the drop in cane yields experienced in the last decade.

Sugarcane is produced in about 100 countries, though a few countries dominate e.g., Brazil, and India, as main producers. It is a difficult crop to harvest since most of the biomass is basically water and fibre. Because of the difficulties associated to harvesting conditions and the relatively small market size, in most countries, when compared to the top world crops[9] (rice, corn, wheat, and soybeans) it is difficult to convince agricultural machine manufacturers to show interest to develop the required technology. For example, lighter machinery, with lower potential to cause soil compaction should be a priority.

8 In 2020, sugarcane average yield for Brazil was 75.6 tons/ha/yr. This yield expresses the harvesting of cane stalks only. If cane straw is added, additional 25-30 tons needs to be added.

9 https://en.wikipedia.org/wiki/List_of_most_valuable_crops_and_livestock_products

There is a considerable potential to grow more sugarcane in Brazil. In a study conducted by Leite et al (2009), to substitute 5% of all gasoline consumed in the world by 2025 (estimated 1.7 trillion litres of gasoline), it would require 102 billion litres of ethanol, considering energy density differences and increased engine efficiencies. The paper identified 12 selected areas with sugarcane potential, located in several states in Central Brazil. These areas were previously occupied with pastureland and considering sugarcane-to-ethanol overall productivity of 6,000 litres/ha/yr[10] using first generation technology, not more than 21 Mha would be required to satisfy the 5% goal. Should this target be achieved, annual ethanol production could be raised from 30 to 102 billion litres, besides the electricity surplus derived from sugarcane bagasse cogeneration. Naturally, there would be positive economic impacts in other agribusiness sectors e.g., yeast production as animal feed.

As feedstock for biofuels, sugarcane presents a very good production model with sugar as coproduct. In the 2020/21 harvest, Brazil produced 41.2 million tons of sugar[11] and exported around 32 million tons[12]) and produced around 29.7 billion litres of ethanol from sugarcane[20]. The reader should be reminded that most ethanol distilleries in Brazil also produce sugar. Around 50% of sucrose entering the mill goes to sugar production (A and B sugars) and the remaining sucrose goes to fermentation for ethanol production. This production model guarantees good profitability since sugar has good international prices. This model was created during the evolution of the ethanol program and is responsible for the elimination of subsidies associated with the ethanol fuel production and use (see Chapter 2).

The so-called "alcohol account" was therefore terminated. Using this model Brazil became the largest producer, at least temporarily,[13] and exporter of ethanol in the world. Unfortunately, sugar market is not as elastic as fuel ethanol. Another consideration is the trend to decrease sugar consumption. So, if fuel ethanol production increases in Brazil, using the same sugar-ethanol model, proportionally more sugar should also be produced. However, world demand for sugar is declining in many countries, primarily in the more advanced economies. Therefore, if Brazil has the intention to expand its ethanol production, another model should be developed or at least complement the existing sugar-ethanol model.

On this regard, although most mills produce sugar and ethanol, they also have capacity to vary the proportion of such products, depending on the economic and technical circumstances. Most mills producing sugar and ethanol have a flexibility level of 60-40, meaning they can adjust the proportion to anything within this range.

According to ANP (2021), in July 2021 the mills authorized to produce ethanol have a production capacity of 130 million litres/d of anhydrous and 230 million litres/d of hydrated. The estimated annual production capacity was 23 billion litres and 41

10 Today the sugarcane ethanol productivity is estimated about 7,000 litres/ha/yr.

11 http://www.iea.sp.gov.br/ftpiea/AIA/AIA-21-2021.pdf

12 https://www.statista.com/statistics/249646/exported-amount-of-sugar-from-brazil/

13 The USA became the largest producer of ethanol in world.

billion litres for anhydrous and hydrated ethanol, respectively. This is based on an average of 180 day harvesting period.

Figure 3.7 below shows historical evolution of bioethanol distilleries from 2006 to 2020 and estimated for 2021 to 2030.

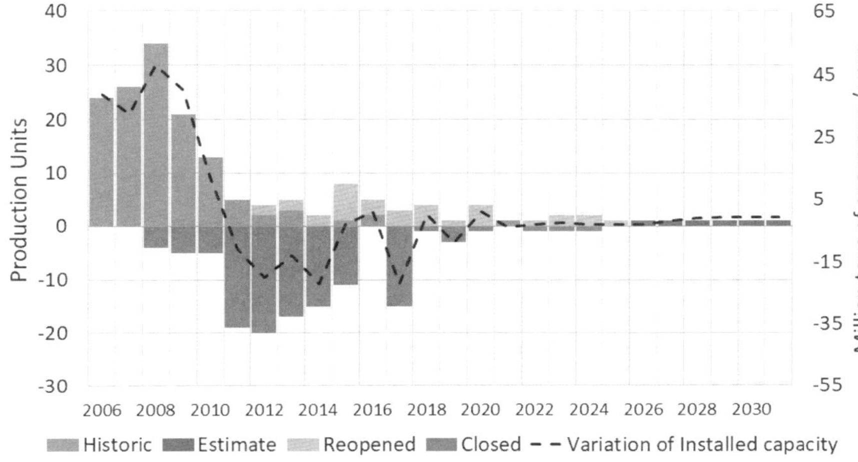

Figure 3.7 Ethanol mills entering in operation 2006-2030 (projected). Source: EPE (2022) based on MAPA, UNICA, and EPE projection.

3.2.3. PRESENT DEMANDS FOR ANHYDROUS AND HYDROUS ETHANOL

The volumes of Otto cycle fuels demanded in the Brazilian market grew steadily from 2008-2013 following closely the economic growth. After this period demand practically stabilized (Figure 3.8).

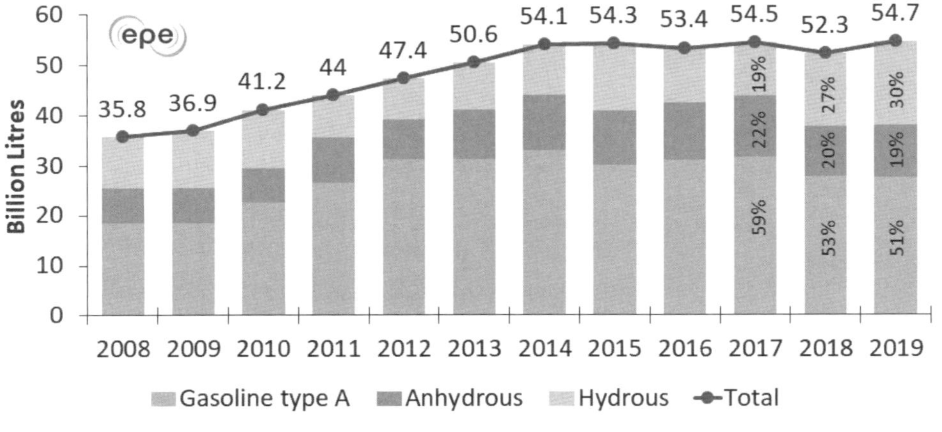

Note: Demand data excludes the CNG share.

Figure 3.8 Otto cycle demand and share of different fuels. Source: EPE (2020).

An important point is the ratio between ethanol price (E100) in relation to gasoline price. Typically, it stays between 65-72% reflecting ethanol lower heating value, as illustrated in Figure 3.9.

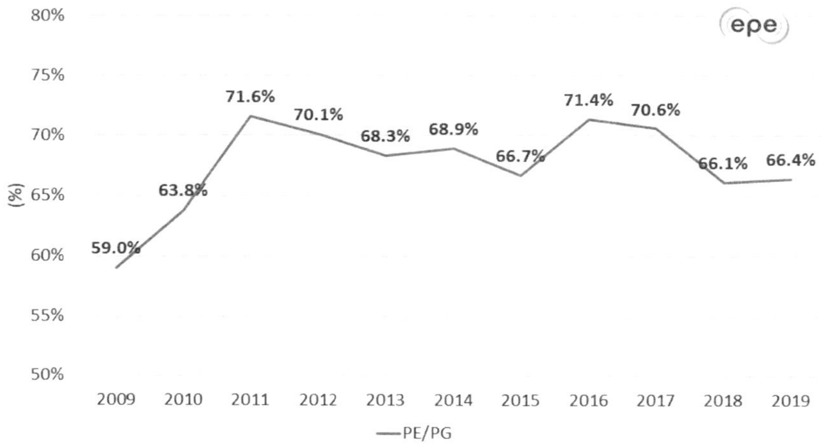

Figure 3.9 Average price of fuel ethanol in relation with gasoline C (E27.5). Source: EPE (2020) from ANP.

3.2.4. PRESENT SITUATION AND MID-TERM POSSIBILITIES OF BIOELECTRICITY IN BRAZIL

Sugarcane bagasse is an excellent by-product. Roughly, about 1/3 of sugarcane are sugars and 2/3 is fibre (bagasse and straw). When we look at these figures it is obvious this crop represents a good agricultural yield, associated with very good potential energy use. A more complete energy breakdown of sugarcane is given below, Figure 3.10.

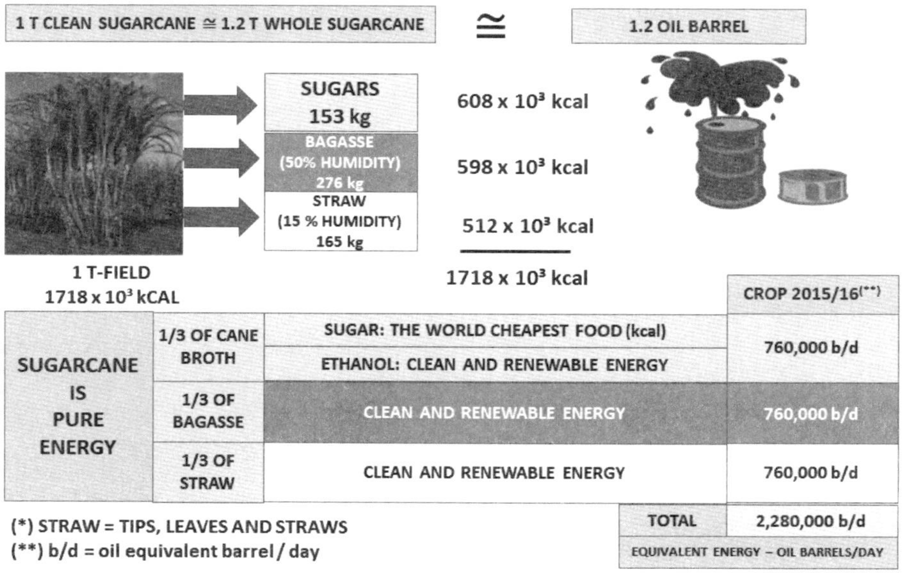

Figure 3.10 Sugarcane composition and its energy value. Source: DEDINI (2016).

The fibres, 2/3 of the cane, are an indication that substantial amount of electricity can be generated from it. In some countries where bioelectricity is highly encouraged such as Mauritius, bioelectricity is sold at premium. Brazil is encouraging its use but still has a considerable potential for improvement.

According to ANEEL (2017) the installed bioelectricity from sugarcane reached 9.4 GW in 2017, an increase of 60% from the previous five years. Of the 378 mills in operation, about 200 sold electricity to the grid and just 50% took active participation in energy auctions. Up to May 2016, 53 auctions took place but just 21 sugar and ethanol mills taking part (EPE, 2022). According to the same source these contracts represented 1.8 GW_{avg} at the end of 2021. Figure 3.11 illustrates well current energy contracted, and the potential expected to be commercialized in the free market, from 2021 to 2025.

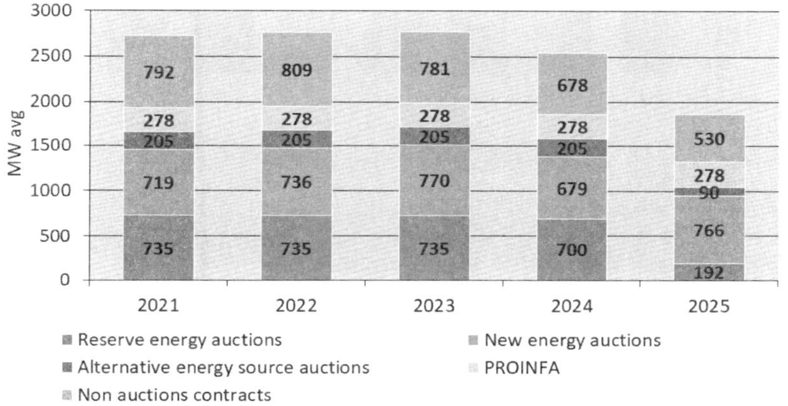

Figure 3.11 Electricity contracted in energy auctions by free market -2021-2025. Source: EPE (2022).

According to EPE the technical potential for selling electricity by the sugar & ethanol mills/distilleries is 6.2 GW_{avg} without considering whole green cane harvesting. If whole green cane was to be considered, using 500 and 787.5 kWh/ton of cane trash and tops respectively, the potential goes up from 6.8 to 10.7 GW_{avg} by 2031 (COGEN, 2009, and EQUIPAV, 2009). Figure 3.12 show current and potential contribution of sugarcane bagasse to the national grid, from 2021 and up to 2031.

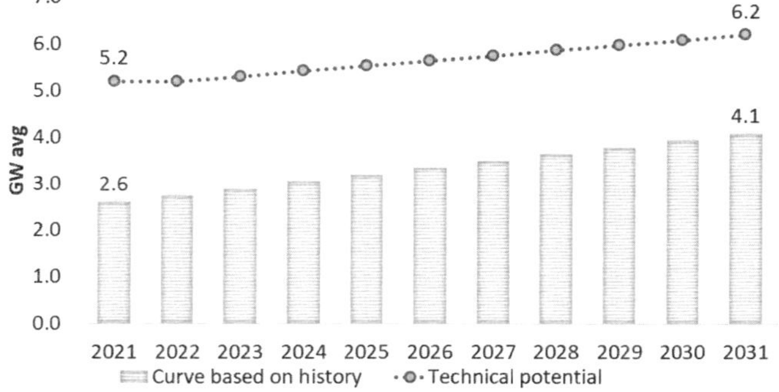

Figure 3.12 Potential for electricity exports from sugarcane bagasse to the grid. Source: EPE (2022).

It is important to mention that the greater share of the total electricity & power generated with sugarcane is consumed inside the mill to provide all energy and heat needs (see Figure 3.13).

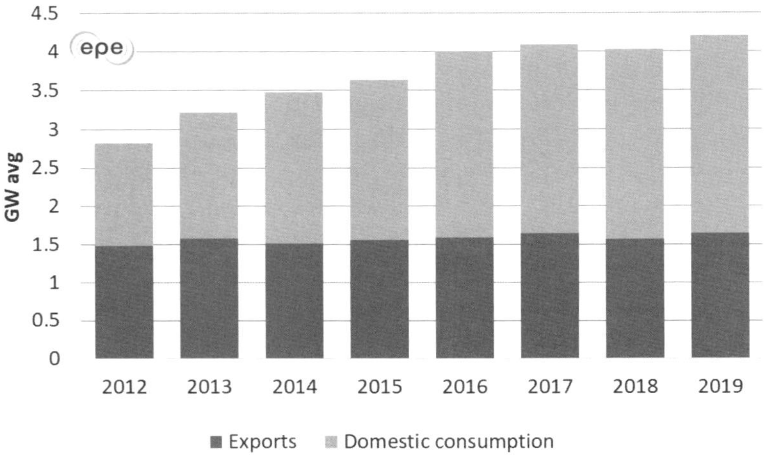

Figure 3.13 Self-consumption and electricity sold by sugar & ethanol mills in Brazil. Source: EPE (2020).

Bagasse is also a good feedstock for G2 ethanol and pyrolysis bio-oil production, as will be discussed further in this chapter.

3.2.5. THE ECONOMICS OF SUGARCANE ETHANOL

Table 3.2 shows the main items integrating the cost of ethanol. It is important to observe the high share represented by feedstock cost, 2/3 of total costs are related to the raw material.

Of the cost of feedstock, harvesting (cutting and loading), and transporting represents almost 50% of overall feedstock costs.

Table 3.2 Raw material, industrial, and administration costs (R$/tc) of sugarcane ethanol in Brazil: harvest 2015/16. Source: Leal (2022) based on PECEGE/I (2015).

	Traditional	Expansion	Northeast
Raw Material	87.42	86.67	105.58
Industrial	28.15	26.73	33.87
Administration	12.78	14.10	16.68
Total	128.34	127.50	156.13

Obs: the expansion areas are mainly concentrated in Central West states of Goiás, Mato Grosso, and Mato Grosso do Sul.

Another important consideration regarding final prices of fuel ethanol in the petrol stations, relates directly to taxes paid by the consumers. Table 3.3 illustrates the importance effects of taxes for both the government in general and PETROBRAS, in particular.

Table 3.3 Composition of gasoline price in Brazil – Feb 2022. Source: LEAL (2022) PETROBRAS, based on ANP and CEPEA/USP data

Price at the Gas Station	R$ 6.79[14]
ICMS (state tax)	R$ 1.75
Distribution and resale	R$ 0.92
Anhydrous ethanol	R$ 0.86
Taxes (CIDE, PIS/PASEP, COFINS)	R$ 0.69
PETROBRAS	R$ 2.37

Transporting ethanol is another important issue, particularly when production plants are located far from consumers. Naturally, less diesel[15] is used for transportation the better the economics and GHG emissions. There are basically two pipelines for ethanol transportation as can be seen in figure below. Figure 3.14 shows the geographical distribution of all ethanol mills in the country (both from sugarcane and corn), and Figure 3.15 the main pipelines. Notice the large concentration in the State of Sao Paulo.

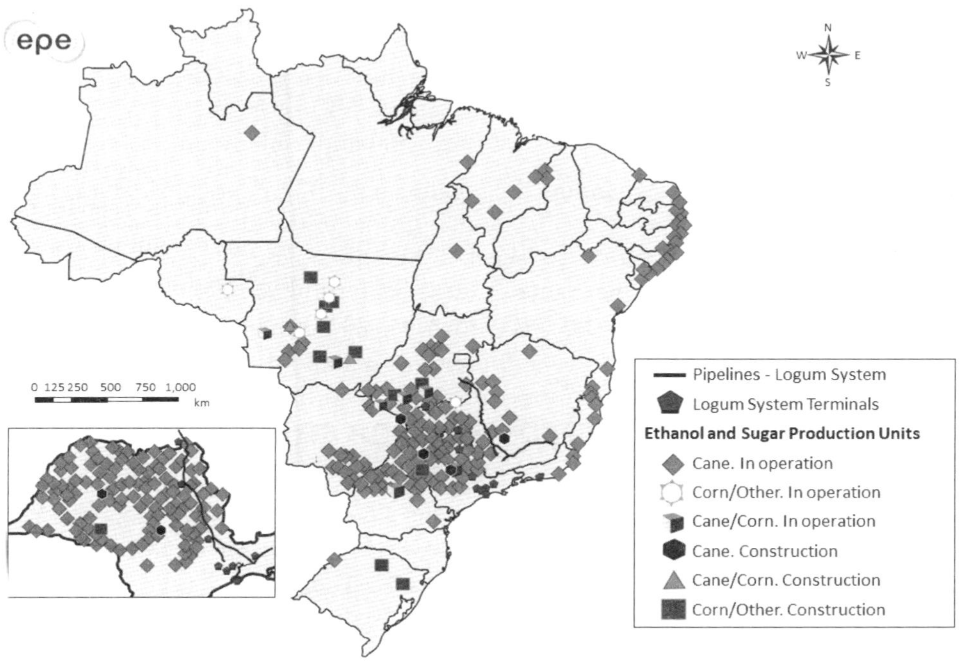

Figure 3.14 Ethanol and sugar producing mills and logistics infrastructure (pipelines) in Brazil. Source: EPE (2022) based on MAPA and LOGUM

14 Exchange rate between R$ 5 to R$ 5.50 during the period.

15 Ethanol is transported by diesel trucks, mainly.

As the market grows more pipelines are being installed. It is important here to remember that these pipes need to be operated throughout the year. Ideally, it would be adequate to have more homogeneous production year-round.

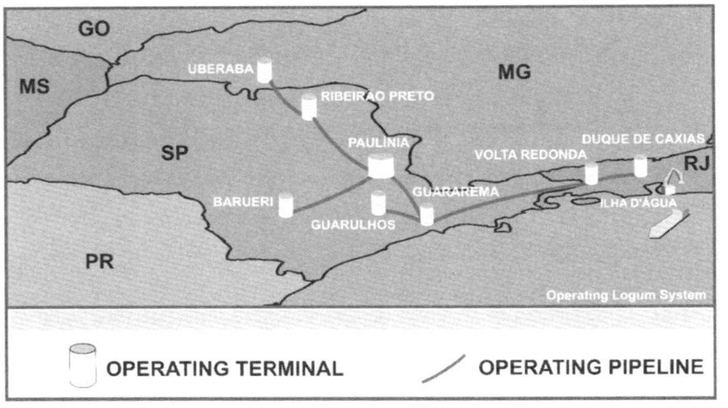

Figure 3.15 Integrated logistics system for ethanol transportation using pipelines in São Paulo and Rio. Source: adapted from LOGUM/EPE (2020)

The pipeline sections that are already in operation are:

- Ribeirão Preto (SP) – Paulínia (operating capacity of 2.8 billion litres/year)
- Uberaba (MG) – Ribeirão Preto (SP) (operating capacity of 1.8 billion litres/year).

In addition, LOGUM also has pipelines such as Paulínia (SP) – Barueri (SP), Paulínia (SP) – Rio de Janeiro (RJ), and Guararema (SP) – Guarulhos (SP). When all these pipelines operate in its full capacity, they will handle 6 billion litres/year.[16]

Among the projects for expanding the pipeline network under consideration, is the planned connection of the Guararema Land Terminal to the distribution bases in São José dos Campos, in São Paulo State. The 42.5-km ethanol pipeline will pass through the municipalities of Guararema, Santa Branca, Jacareí and São José dos Campos with a yearly capacity of 40 million litres (NOVACANA, 2020).

3.3. CORN: THE NEW FEEDSTOCK FOR ETHANOL IN BRAZIL

Corn is also a traditional crop in Brazil and planted in two different seasons: summer (as a main crop harvest or "*safra*") and winter (as second crop or "*safrinha*", after soybeans is harvested). There is also corn harvested as third crop in Northeast Brazil, although the production is relatively small (MACHADO, 2021). As can be seen, most of the corn produced today, around 70-80%, is as second crop sharing annually land with soybeans (Figure 3.16).

16 https://www.udop.com.br/noticia/2021/08/05/dutos-da-logum-chegam-a-grande-sao-paulo.html

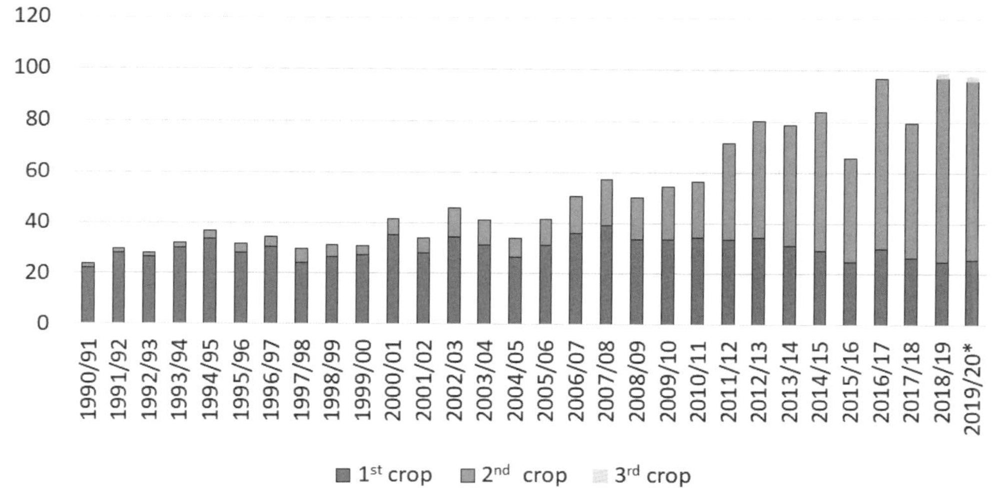

Figure 3.16 Overall corn production in Brazil. Source: CONAB / elaborated by Scot Consultoria.

Ethanol using corn as feedstock is growing rapidly in Brazil, as shown in the figure below.

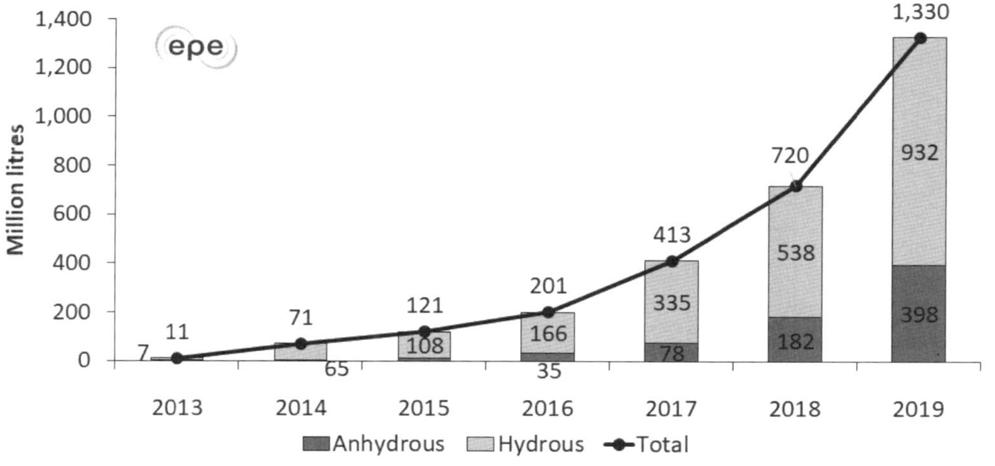

Figure 3.17 Corn ethanol production in Brazil. Source: EPE (2020) from UNICA.

Currently about 19.9 Mha are devoted to corn in Brazil, with a production of 87 Mt in 2020/21; it is expected to reach 20.9 Mha and 112.9 Mt in 2021 to 2023 (CONAB, 2022). It is important to emphasize that "safrinha" uses the same agricultural land of corn and soybeans. The main idea is to use the remaining fertilizer left after soybeans are harvested, as it is well known their nitrogen fixation effect.

Geographically, corn is planted in several states, but mostly in Mato Grosso and Paraná. Because of the nature of corn, it facilitates its cultivation in different climate such as Central-West and South of Brazil.

Presently, corn production supplies three market objectives: the domestic market, exports, and ethanol production. Brazil is the world's second corn exporter with nearly 10-15%.[17] Around 3.7 billion litres of ethanol were produced in 21 plants (both corn dedicated and flex), predominantly in Mato Grosso and Goiás states; all ethanol production uses corn as feedstock. It is predicted that corn ethanol could reach 9.0 billion litres/year in 2031 in 54 ethanol plants, as shown in Table 3.4.

Table 3.4 Installed capacity of corn ethanol plants in Brazil. Source EPE (2022) based on MAPA (2021)

	2021		2031	
	Number of units	**Installed capacity of ethanol (billion litres)**	**Number of units**	**Installed capacity of ethanol (billion litres)**
Total	21	3.7	54	9.0
Flex	11	1.2	20	2.9
Full	100	2.5	34	6.1

The prospects of corn ethanol production in Brazil are very good. The life cycle of corn ethanol is also good because other sources can be used as an energy source e.g., eucalyptus and sugarcane bagasse, in corn ethanol plants, particularly when there is a possibility to integrate corn and sugarcane production in the same plant. There are some other benefits associated e.g., continuous operation (all year around) of the plant without the need for storage tanks since the off-season[18] period can be eliminated.

3.4. THE PRESENT STATUS OF BIODIESEL IN BRAZIL

The world production of biodiesel in 2020 was approx. 47 billion litres (FAME and HVO), and Brazil contributed with about 6.4 billion (~14%), being the 3rd largest producer behind Indonesia (~17%) and the USA (~15%) (REN21, 2021). (See also Chapter 1).

Originally, when PROBIODIESEL was setup and during the early years of its implementation, the idea was to use small family farms and alternative feedstocks such as sunflower and castor oil. This was not a realistic approach. As explained in Chapter 2, the reasons why sugarcane succeeded as feedstock for ethanol production was the existence of expertise and knowledge in sugarcane cultivating in large scale, in addition to a well-developed infrastructure and the influence of the sugarcane lobby. This lesson was ignored when PROBIODIESEL was first set up. Naturally, the most important crop in Brazil, soybean, became the best candidate to produce large scale vegetable oil for biodiesel production. Soybeans is a seasonal crop, typically planted with the

17 https://www.gov.br/pt-br/noticias/agricultura-e-pecuaria/2021/10/producao-de-etanol-de-milho--avanca-no-pais-como-opcao-sustentavel-e-de-valor-agregado

18 Sugarcane ethanol plants in Brazil are highly concentrated the Central South region (about 80% of the country´s production. Ethanol production in the Central South region is normally done in between April and October.

first rains in October and harvested near April. Soybeans, shares the same land with "safrinha" corn, so if there is rainfall delay this may affect the "safrinha" corn planting and productivity. Also, soybeans are crushed to produce soybeans oil, by far the least expensive cooking oil in Brazil. In the crushing process, soybeans cake is produced, which is mostly used as animal feed primarily in poultry and pigs' farms.

Brazil is the world's largest producer and player in soybean grains exports (IBGE, 2021). Although soybean was not an energy crop, Brazil had sufficient knowledge on how to cultivate it in large commercial scale; there is a good infrastructure and well-organized and powerful producers' associations. The Brazilian Association of Vegetable Oil Industries (ABIOVE) is the largest private group representing the soybeans sector.

Another sector that benefited from large scale biodiesel production is the beef cattle industry. Tallow is produced in large quantities and so far, there was not a good market for it.

Other raw materials such as cotton, corn and residual oils also benefited from PROBIODIESEL because of its regional advantages, seasonal availability, and share logistics. Currently, a large spectrum of raw materials is used to produce biodiesel, although nearly 68% is from soybeans (Figure 3.19 below).

Soybean requires substantial amount of land, and hence is not as good energy crop as sugarcane. This results in low oil productivity (about 600 litres of soybean oil/ha/yr), compared with palm oil in Malaysia with about 5,000 litres of oil/ha/yr. However, it is not foreseen soybeans will lose its leadership as raw material for biodiesel in Brazil, at least not until 2030. It would be important to encourage other alternative energy crops such as palm oil to grow for this important and growing market.

Few palm oil projects were established in Brazil, with AGROPALMA being one of the most important. Also, the iron ore company, VALE, is involved in setting up palm oil plantations. For example, it has set up BIOPALMA in Pará state, where it has a 57,000 ha of palm plantations.[19] The main difficulties so far have been associated with agronomy aspects.

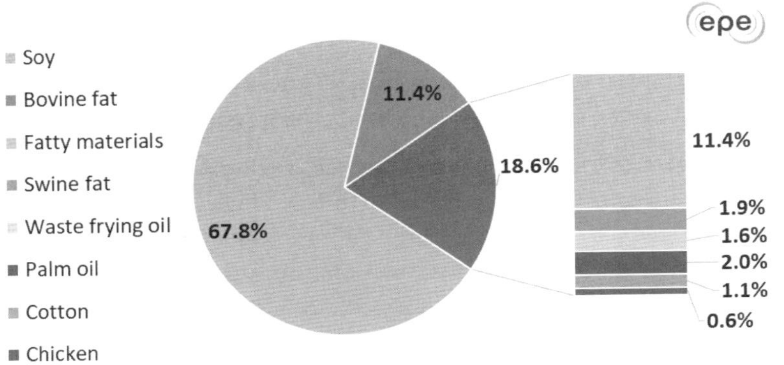

Figure 3.18 Share of raw materials for biodiesel production in 2019 in Brazil. Source: EPE (2020) from ANP (2020).

19 https://ubrabio.com.br/2013/05/22/vale-investimente-na-producao-de-oleo-de-palma/

If Brazil wants to export less soybean grains and process them domestically benefiting the production of vegetable oil and soybean cake, it should change the existing exemption of export taxes given to raw materials.[20]

Production of soybean and exports continue to increase. According to IBGE (2021) soybean production went from 128 million tons in 2020 to 133 million tons in 2021. Soybeans grains exports increase to 83 million tons in 2020 (12% higher than 2019) (ABIOVE, 2021a).

As for the production costs associated with biodiesel, the raw material typically represents about 80% (IEA, 2004). EPE studies indicate there is a correlation between prices of biodiesel and prices of soybean grains and soybean oil in the domestic market.

It is expected that during the next 10 years biodiesel demand will practically double (Figure 3.19).

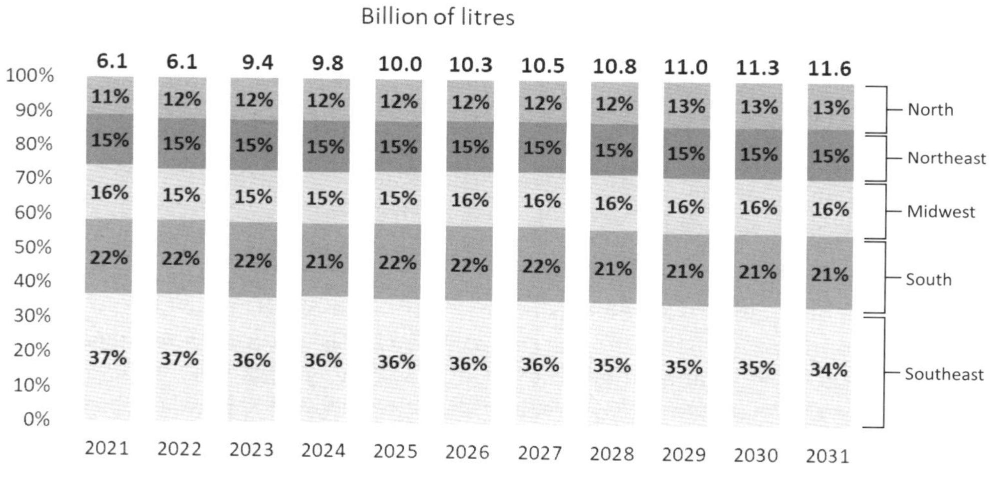

Figure 3.19 Biodiesel expected demand by Brazil. Source: EPE (2022).

It is also Important to note from the Figure 3.19 above, that there is a certain equilibrium in the production of biodiesel from different regions, although demand is different.

Table 3.5 summarizes the expected biodiesel processing capacity and compulsory consumption for different regions, from 2020 to 2031. Figure 3.20 shows the map where biodiesel plants are located which, as can be appreciated, Mato Grosso has the largest concentration. And Figure 3.21 summarizes blending and the regulatory framework. As can be seen blending increased from 2% in 2005 to an expected 15% in 2023.

20 The Kandir Law in Brazil, exempts raw materials such as soybean grains on paying ICMS taxes for exports.

Table 3.5 Expected biodiesel processing capacity and compulsory consumption for different regions in Brazil from 2020 to 2031. Source: EPE (2022).

Region	2021			2031		
	Nominal installed capacity	**Compulsory consumption**	**Balance**	**Nominal installed capacity**	**Compulsory consumption**	**Balance**
North	320	756	-436	932	1,525	–593
Northeast	880	971	-91	1,240	1,791	–551
South	4,569	2,437	–1,613	5,057	2,443	–1,313
Southeast	824	1,413	3,156	1,130	3,978	1,079
Central-West	4,728	1,031	3,697	5,932	1,818	4,114
Brazil	11,321	6,609	4,712	14,291	11,557	2,734

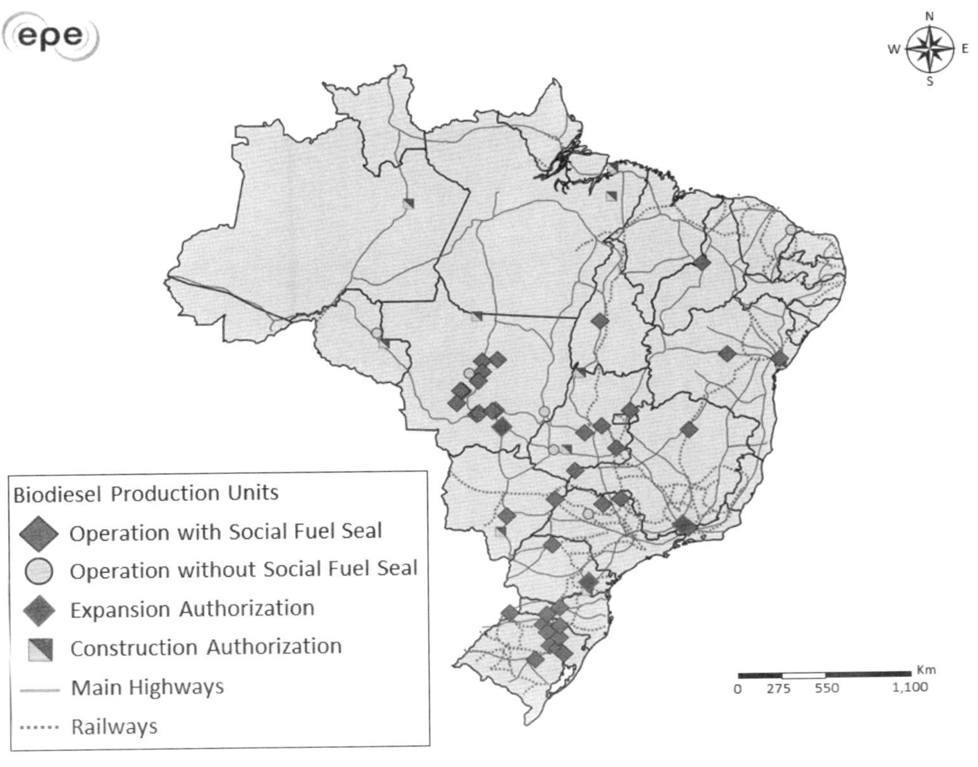

Figure 3.20 Localization of biodiesel plants and its logistics routes. Source: EPE (2022).

The evolution of biodiesel mandates is given in the figure below.

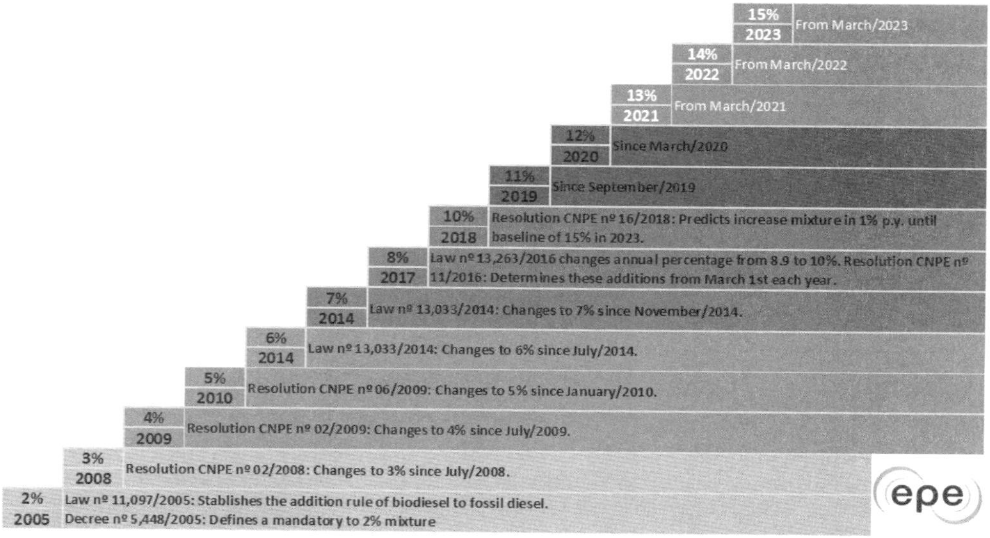

Figure 3.21 Evolution of biodiesel regulatory framework in Brazil. Source: EPE (2020).

It is also important to observe biodiesel prices volatility which oscillate significantly, probably in response to soybean prices and its costs, as shown in Fig. 3.22.

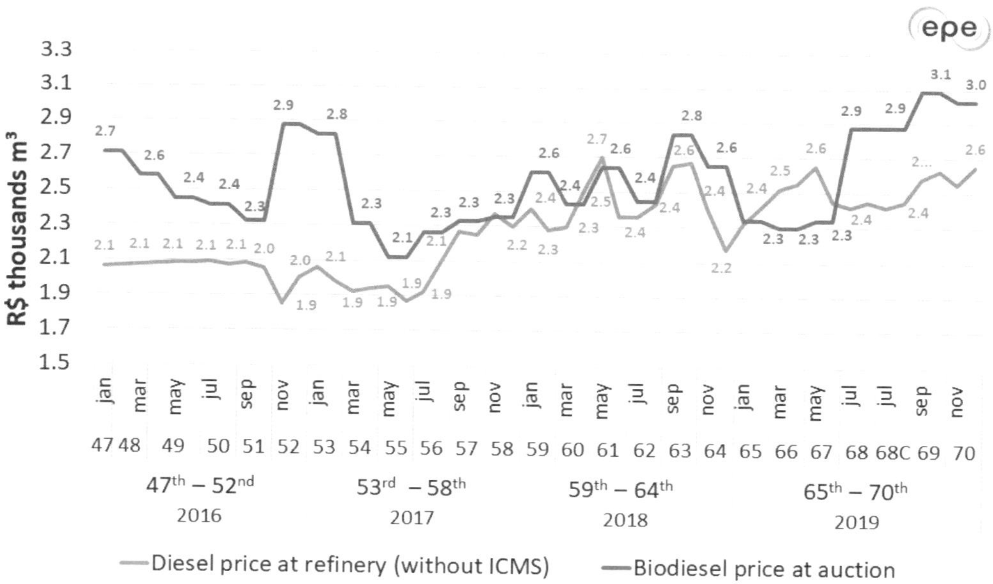

Note 1: Biodiesel prices corresponds to the indicated auctions.
Note 2: The price of diesel corresponds to its value at the refinery.
Note 3: Diesel and biodiesel prices are presented in current values.

Figure 3.22 Average prices for biodiesel and diesel in Brazil, without taxes ICMS), from 2016 to 2019. Source: EPE (2020).

3.5. CAN EUCALYPTUS BE A FEEDSTOCK FOR BIOFUELS?

Eucalyptus was introduced in Brazil in the mid-20th century. Under Brazilian edaphoclimatic conditions this fast-growing tree adapted very well in the country and contributed to the production of cellulose and multi-use construction wood. Additionally, eucalyptus helped the reduction of unnecessary deforestation caused by charcoal production used in the pig-iron steel industries (ROSILLO-CALLE et al, 1996).

Currently eucalyptus plantations occupy approximately 10 Mha, notably in the states of Minas Gerais, Espírito Santo, Pará, and Bahia. Eucalyptus is a permanent crop with a short production cycle compared to other tree species. After the plant is stablished, the first cut should be anything between 5 to 7 years, depending on its final use. Also, the culture is very adapted to hilly conditions as it happens in Central states and therefore eucalyptus do not compete directly with food crops such as soybeans, corn, or even sugarcane, since they require different type of mechanization. It helps that about 30% of Brazilian territory is hilly and therefore unsuitable for mechanization.

Finally, eucalyptus is more flexible with regards to water requirement. It is much more resilient than other plants when periods of drought occur, as it is happening periodicity in Central Brazil, notably in the last two decades.

As stated before, eucalyptus is also a good candidate for bioelectricity, charcoal, and/or bio-oil production, which are been considered for maritime transport. Brazil should seriously consider the use of eucalyptus to increase the resilience in projects concerning corn ethanol (energy supply), sugarcane ethanol (electricity exports), and production of biofuels for other new markets such as maritime transport.

3.6. SOLID URBAN WASTES (SUW) AND ITS POTENTIAL IN BRAZIL

Solid wastes are not the most suitable source for biofuels production, but its use is important for GHG emissions and sustainability criteria. This worldwide abundant feedstock is in fact the result of several wasteful inefficiencies, particularly in large cities of middle income and poor countries. Brazil is not an exception, since nearly 80% of the country population live in urban areas today and urban wastes represents a serious environmental problem.

SUW is beginning to be used in Brazil, particularly for recovering methane produced in landfills e.g., in the Bandeirantes landfill nearby São Paulo where a 20 MW thermoelectric plant was installed for that purpose.[21]

With the objective to regulate this sector, the Brazilian federal government, through IBAMA, approved in 2010, *Law n. 12.305* creating a national policy of solid wastes.[22]

21 https://www.infraestruturameioambiente.sp.gov.br/noticias/2004/03/termeletrica-do-aterro-bandeirantes-reduz-emissao-de-co2-para-atmosfera/

22 https://antigo.mma.gov.br/cidades-sustentaveis/residuos-solidos/politica-nacional-de-residuos-solidos.html

3.7. LAND USE OPPORTUNITIES AND BIOFUELS PRODUCTION

At this point we may ask, what land use has to do with biofuels in Brazil given its enormous land mass? What is the possible relationship and implications?

When the Environmental Protection Agency (EPA)[23] of the USA classified biofuels according to their GHG emissions mitigation potential, they stated that Brazilian sugarcane ethanol has the potential to reduce up to 61% GHG emissions when compared to gasoline. This was somewhat 20% lower than expected as calculated by the Brazilian experts. Why this discrepancy? Because EPA introduced a new iLUC concept (or indirect land use change) reducing the mitigation potential.

Direct land use change (LUC) is the direct change impact in the use of land. So, when 1 ha of pasture is replaced by sugarcane, the LUC effect is the GHG emissions accounting of that 1 ha going from pasture to sugarcane. The iLUC is a more sophisticated concept in which it takes also into consideration the possibility of that 1 ha of pasture, reappears when deforestation occurs in the Amazon Forest, for example. So, 1 new ha of sugarcane not only occupies the 1 ha of pastureland, but it also destroys 1 ha of forest, through deforestation. So, both effects LUC and iLUC should be considered when calculating the life cycle of ethanol though sugarcane expansion.

This is, of course, a controversial subject. It is very difficult to establish this direct relationship. Most likely the deforestation process in Brazil has much more to do with other actions and reasons, such as former governmental policies, such as INCRA, to promote land occupation in the agricultural frontiers, promoting the movement of people to occupy new lands, fiscal incentives, illegal exploitations, etc. In other words, sugarcane and biofuels are not responsible for additional deforestation.

Whatever reason, the fact is that many researchers[24] believe in iLUC, and further research needs to be done to clarify this issue, back up by scientific data. With 1.1% of Brazilian territory occupied by sugarcane, the country is number one as sugar producer and exporter, and second as fuel ethanol producer. Ethanol is almost a coproduct of sugar, and it can be said that one benefits the other from the economic and environmental points of view.

The problem is that this nice synergy between sugar-ethanol can no longer adequately reflects increase of ethanol production. There is not a worldwide market for sugar, should ethanol production increase significantly, and hence the parity sugar-ethanol model needs to be changed.

So, the question is, how can fuel ethanol production increase significantly and sustainably from an economic point of view? It seems clear that a new model is therefore needed.

There are two possibilities: 2nd G ethanol or corn ethanol. 2nd G ethanol from sugarcane is expected to mitigate 81% of GHG emissions. Corn ethanol produced in Brazil

23 http://www.ethanolproducer.com/articles/15755/increasing-the-role-of-advanced-biofuels-in-the-us

24 https://en.wikipedia.org/wiki/Indirect_land_use_change_impacts_of_biofuels

can offer different opportunities. Both possibilities are feasible, and one complements the other. Both represent a change of existing model and should be encouraged.

This is, of course, good news. However, we need to move faster and look for different production models, such as the American model, to produce ethanol and beef. The US corn ethanol has a relatively bad reputation as it has lower productivity when compared with sugarcane ethanol[25] and also, much lower GHG emissions mitigation potential than sugarcane.[26]

The USA has developed a very solid system for producing ethanol from corn while using its main by-products DDG or WDG.[27] Both, the high protein DDG and WDG, can be used as animal feed, typically cattle. This system is efficient because it allows a significant lower use of land than natural extensive pasture.

About 25 years ago, Brazil was a good example for producing beef in grazing and natural pastureland because the world was seriously affected by the mad cow decease. However, this situation has changed, particularly in the last 15 years, when Brazilian sugarcane ethanol was also criticized because of the iLUC effect.

Brazil dedicates around 20% of its 850-855 Mha to pastureland, with an average animal density of about 1 animal/ha. However, it can vary from anything below 0.5 to 3 or 4 animals/ha. Pastureland supporting capacity depends very much on several factors such as how cattle and land is managed,[28] quality of land and rainfall, etc., but it can certainly be improved.

Therefore, if the animal density is raised from 1 to 2 animals/ha, then a significant amount of land (nearly 100 Mha) could become idle/spared. This increase in animal density could be accomplished by implementing actions, such as:

- Creating a national and effective policy on *Land Use Reform*, to encourage better land occupation while protecting ecological sanctuaries such as the Amazon Forest and Pantanal.
- Introducing corn ethanol in combination with cattle integration, particularly in areas and states where pastureland is presently occupied with low animal density.

A high priority should be given to the southern border of the Amazon region, the Pantanal, and few other regions with more intensive cattle production. Those are endangered regions and should be prioritized and protected.

Cattle-ranching is still a highly extensive activity in Brazil with relatively small percentage of intensive farms e.g., approximately 5%. What is proposed here is not

25 Sugarcane ethanol produces around 7,000 litres/ha/yr and corn ethanol around 4,000 litres/ha/yr

26 Corn ethanol typically uses natural gas as a source of energy and sugarcane ethanol uses sugarcane bagasse. This difference explains the between life cycle of sugarcane ethanol.

27 In the USA there are two methods to produce corn ethanol, the dry method that generates the Dry Distillers Grains – DDG and the wet method that generates the Wet Distillers Grains – WDG.

28 Pastureland support capacity certainly depends on climatic conditions which may vary seasonally.

necessary to go from extensive to highly intensive cattle production. What is proposed is to go from highly extensive to semi-intensive cattle production as a mechanism to free land for other uses, including the regeneration of the native forests.

This *Land Use Reform* could result in the following benefits:

- Help to protect endangered ecological sanctuaries
- Allowing more sustainable agriculture to expand to pastureland
- Increase cattle production sustainably
- Increase biofuels production in combination with food and avoiding any iLUC effect.

There are still few important questions, requiring answers e.g.,

- How to adapt existing cattle breeds to a more intensive production?
- What is the DDG/WDG "shrinking" land capacity?
- What would be the new corn ethanol life cycle under these new conditions?

3.8. IMPACTS: AGRICULTURE, SOCIAL, AND ENVIRONMENT

The initial objective to produce biofuels locally to enhance energy security in Brazil has been fully accomplished. Now, the intention is to expand biofuels production to contribute to enhance sustainable development.

As far as GHG emissions, Brazil has a peculiar situation. Most of its emissions are non-energy related, as it happens in most of large economies, and relate either to land-use and forestry or agriculture. Energy is responsible for about 1/5 of the country total GHG emissions. So, the first conclusion derived from these data is that, if something is to be done to reduce GHG emissions, the focus should be first on land-use and forestry, and secondly in agriculture and then energy (see Table below).

Table 3.6 Brazil´s emissions in 2019 and 2020 (tCO2e – GWP-AR5). Source: SEEG (2021).

Sector	2019	%	2020	%
Agriculture	562,987,702	29%	577,022,998	27%
Energy	412,466,747	21%	393,705,260	18%
Industrial processes	99,472,616	5%	99,964,389	5%
Waste	90,399,714	5%	92,047,812	4%
Land-use change and forestry	806,996,124	41%	997,923,296	46%
Total gross emissions	1,972,322,903		2,160,663,755	
Total net emissions	1,336,613,309		1,524,954,161	

When examining Fig. 3. 23 below, it is important to recognize that significant reductions can be observed on land-use and forestry between 2003 and 2009, probably because of economic development downturn during that period.

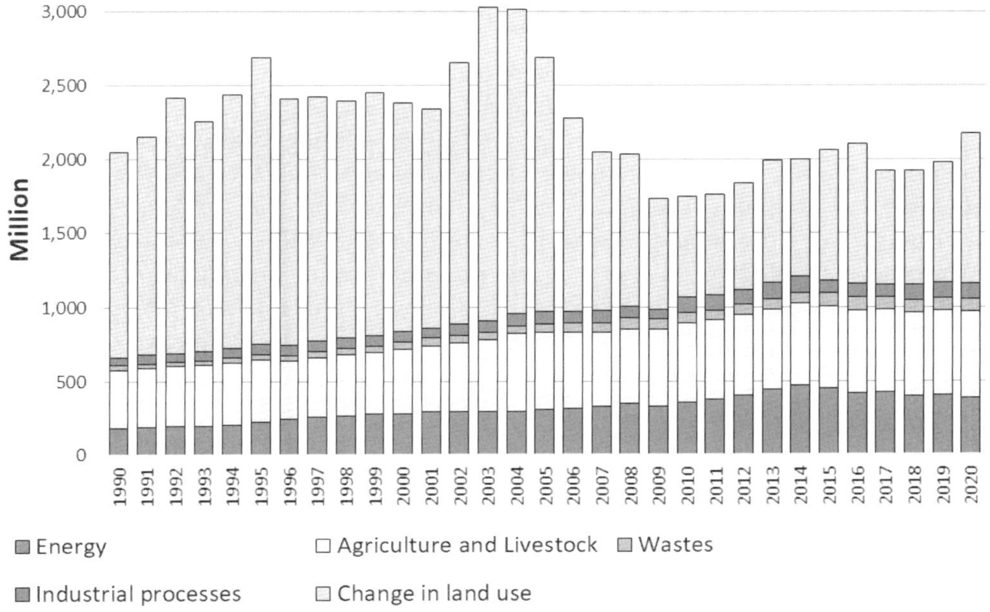

Figure 3.23 GHG emissions in Brazil from 1990 to 2020 ($GtCO_2e$). Source: SEEG (2021).

However, in recent years, due to the economic and political crises, there was a deterioration in controlling deforestation. Also, because the Federal Government has relaxed deforestation control.

3.9. BIOFUELS TECHNOLOGICAL INNOVATION FROM AGRICULTURE, PROCESSING TO FINAL USE

Innovation on biofuels needs to be understood as a series of R&D&I actions from agriculture to processing and final use (engines). As mentioned in the "lessons from biofuels" section, to accomplish an overall high efficiency, biofuels need to constantly pursue improvements in agriculture, processing, and final use, at the same time.

Throughout the implementation of Brazilian biofuel programs, both PROALCOOL and PROBIODIESEL, there was a constant preoccupation with technological innovation. Both programs also benefited from R&DD policies created during the 1990s when the energy agencies, ANP for petroleum and gas, and ANEEL for electricity, were created. With the setup of ANP,[29] which today also involves biofuels, and

29 Agência Nacional do Petróleo, Gás Natural e Biocombustíveis -ANP (National Petroleum Agency).

ANEEL[30] specific taxes were directed to promote technological innovation ranging from new processes to greater operational and energy efficiency.

3.9.1. INNOVATION IN AGRICULTURE

A considerable effort has been devoted to improving existing cane variety programs e.g., PLANALSUCAR, IAC, and COPERSUCAR. PLANALSUCAR, today RIDESA, was a program aimed to produce better varieties considering different edaphoclimatic conditions present in different areas of Brazil. It can be said that it is the most efficient national programs. The IAC is a State of São Paulo program to develop new and improve varieties, focusing on Southeast conditions. COPERSUCAR, today Centre of Sugarcane Technology (CTC), is a private program producing varieties for its cooperative members. It can be stated that there is a good complementarity between these three programs which have been quite successful in improving productivity for sugarcane (GAZZAFFI et al, 2010).

When PROALCOOL started in 1975, there was a huge concern with the problem posed by large volumes of vinasse. This limitation was first considered as a considerable obstacle to large-scale ethanol production from sugarcane. However, here was an important innovation on potassium-rich vinasse disposal as fertilizer thanks to scientific research and collaboration between the private sector and universities (MUTTON et al., 2010). Although it should be added that vinasse disposal still requires more attention. It is a high-volume residue from fermented sugarcane juice disposed in different ways among sugarcane ethanol producing countries, e.g., India and Colombia. Currently considerable progress has been achieved with vinasse biodigestion, as it will be discussed further bellow. A law was promulgated by the Secretary of Environment of the State of São Paulo, José Goldemberg, early this century, to promote green sugarcane harvesting in São Paulo State. This law allowed sugarcane growers a transition period of various years to replace burning to non-burning harvesting (green harvesting), so that the necessary investments, adjustments, and implementation, etc., could be made.

After the introduction of sugarcane mechanical harvesting there was some stagnation in productivity probably due to two phenomena: harvesting equipment, machinery, and trucks, were much too heavy and less precise cutting than manual labour, as already discussed. The long-term effect, associated with adverse climatic conditions, and lack of adequate investment particularly on fertilizers, has resulted in sugarcane productivity stagnating in the last 20 years in the country.

A positive impact of the non-burning law was an environmental improvement in the rural e areas e.g., improved air quality during the winter months. Heavy machinery also caused soil compaction and sugarcane root system bruising and more dirt to the mills. Naturally, another benefit was the resulting straw which increased the quantity of recoverable fibres but also worsen the cane cleaning process at the mill. From the

30 Agência Nacional de Energia Elétrica – ANEEL (National Agency of Electric Energy)

agronomic and industrial points of view, the introduction of machinery resulted in the deterioration of cane quality and increased costs.

The introduction of cane harvesting machinery, with corresponding need to improve harvesting synchronism and organization, favoured the introduction of information technologies. Nowadays the sugarcane sector uses IT allowing feedstock cost reduction and more effective and sustainable production monitoring.

There were also other improvements i.e., cane planting. Traditionally cane stalks were planted manually, which was substituted by "Plene", a technology developed by Syngenta to modernize cane planting. Another company, EACEA, also revolutionized sugarcane planting by producing-Sprouted Seedlings (MPB) in high-tech agricultural greenhouses installed next to ethanol distilleries. With operational costs mitigated through the recovery of tailings and CO_2 from the distillery, the high-quality seedlings cut prices compared to traditional planting, in a market estimated to be worth 7 billion reais (nearly US$ 1.3 billion). The sugarcane planting model through MPB changed 300 years of planting practices, by freeing up millions of tons of cane to produce sugar or ethanol, and even allowing for the immediate introduction of new, more productive, and more resistant varieties to climatic rigors and losses from diseases and pests. Figure 3.24 shows sugarcane seedlings grow in greenhouses.

Figure 3.24 Sugarcane seedlings been grown in greenhouses as part of new planting systems. Source: image provided by Andrés da Silva (EACEA).

3.9.2. INNOVATION IN PROCESSING

An area which received special attention, notably after the 2002 energy crisis, was bioelectricity from sugarcane. The sugarcane sector was requested to contribute to the national effort to diversify energy sources, too dependent on hydro power. An important study was carried out by Sousa and Macedo (2010) to assess the potential contribution of sugarcane bioelectricity. A more recent study was carried out by Leal (2020) on sugarcane renewable electricity, including sugarcane straw and other possibilities. The project estimated that sugarcane primary energy is approximately 600 GJ/ha (nearly 2/3 is fibre) or nearly 100 barrels of oil.[31] So, in theory, annual fibre production from sugarcane represents nearly 1.8 million barrels/day.[32] Therefore, substantial surplus bioelectricity could be produced by the mills, particularly with higher prices if so required. Improvements in energy efficiency could result in even greater optimization of resources e.g., in boiler efficiency.

In the processing area, there is also a constant improvement such as equipment, processes, and practices. Besides improvements to allow greater electricity surplus, there was an also increasing concern to improve sustainability indicators. Significant achievements were obtained particularly on by-products production and applications, such as the DEDINI "zero effluents" concept (OLIVÉRIO, 2014). This concept allowed the development of several new products using just by-products.

Important breakthroughs have also been made on 2G ethanol production in the last 15 years. Progresses have been both in the scientific and entrepreneurial areas. For example, various research centres have been created e.g., Embrapa Agroenergia in 2006, and CTBE, today LNBR in 2010, and dedicated specific research programs such as FAPESP/Bioen in 2008.

In addition, BNDES created in 2011, PAISS[33] innovation program to encourage 2G ethanol plants. As a result, several projects and partnerships have been set up. For example, GRANBIO started operation of its Bioflex plant in 2014 in Alagoas, with a capacity of 84 million litres/year of 2nd G sugarcane ethanol. Today it produces around 30 million litres/year. RAÍZEN from Shell/COSAN, in Piracicaba, produces 41 million litres/year.[34] The new RAÍZEN G2 ethanol plant, due to start operation in 2023, will produce 82 million litres/year. According to these sources, together GRANBIO and RAÍZEN will produce around 150 million litres in 2023.

More recently, biogas production from vinasse biodigestion has received additional incentives. Vinasse application as fertilizer liberates methane. To avoid this unnecessary methane emissions, vinasse can be biodigested and its resulting methane use to produce electricity reducing its atmospheric impact and still making possible effluent use as fertilizer and reduce costs.

All these accumulated improvements on sugarcane agriculture and processing, led to a "learning curve" already mentioned in Chapter 2. This learning process brought cost reduction and improvement in sustainability indexes, making sugarcane ethanol more competitive with gasoline (GOLDEMBERG, 2004).

---

31 Primary energy of 1 barrel of crude oil is approximately 6.1 GJ.

32 Estimated using MME (2021) data.

33 https://www.bndes.gov.br/wps/portal/site/home/imprensa/noticias/conteudo/20130912_raizen

34 https://www.energiaquefalacomvoce.com.br/2021/06/30/fns-brasil-se-prepara-para-ofertar-150-milhoes-de-litros-anuais-de-etanol-2g-e-a-maioria-ja-esta-vendida/

FINAL REMARKS

Several energy crises have affected Brazil in the last 100 years with important economic and political impact. Typically, all energy crises reflected higher oil or electricity prices. These increases in energy prices somehow created the political will and the economic conditions to encourage RE, in particular biofuels. Since biofuels, typically need either subsidies or a well-structured economics, the raise of oil prices accomplished this task.

The present situation of renewable energies, in particular biofuels (ethanol and biodiesel) in Brazil, is benefiting from present high cost in oil prices, in a moment when the world the economic situation is worrying. Other factors such as climate change, the raise of China, and now the pandemics and the Russian war on Ukraine, are major new challenges for biofuels.

Brazil has been the subject of world attention particularly when it comes to the Amazon Forest. The deforestation of tropical forests affects both climate and global warming, and Brazil is considered the most important emitter of GHG emissions from deforestation. In the last two decades sugarcane ethanol and its iLUC was blamed for deforestation when in fact the culprit has been extensive cattle-ranching and soybean expansion. On the other hand, sugarcane ethanol currently faces a considerable challenge because of a combination of difficulties, including:

- Expanding production. In the present model sugarcane ethanol is only viable if its economics is compensated and supplemented with sugar sales
- Electric mobility. This option has been presented as a "clean" solution, although the main issue is how much sustainable electricity is going to be generated.
- Domestic difficulties associated with relatively low engine efficiencies. Flex-fuel engines are no longer adequately satisfying consumer needs.

As biofuels in Brazil do contribute significantly to mitigate GHG emissions, it is imperative to discuss new production models to satisfy the present needs, and to connect these modern biofuels with modified improved engines. These solutions require mid and long-term policies involving not only Brazil but also private enterprises and international cooperation.

REFERENCES

ANP – Agência Nacional do Petróleo, Gás Natural e Biocombustíveis. (2021). *Autorização para produção de biocombustíveis*. Rio de Janeiro. Available at https://www.gov.br/anp/pt-br/assuntos/producao-e-fornecimento-de-biocombustiveis/autorizacao-para-producao-de-biocombustiveis.

COGEN. (2009). *Bioeletricidade* – Reduzindo emissões e agregando valor ao sistema elétrico. Associação da Indústria de Cogeração de Energia.

CONAB – Companhia Nacional de Abastecimento. (2022). *Séries históricas*. Cana-de-açúcar, CONAB, Brasília. Available at https://www.conab.gov.br/info-agro/safras/serie-historica-das-safras?start=20

DEDINI. (2016). *A DEDINI e o Esforço Privado nestes 40 Anos de PROÁLCOOL, do engenho de açúcar à usina sustentável!!!* São Paulo, 30 November 2016. PPT presentation. Available at https://fapesp.br/eventos/2016/11/proalcool/Dedini.pdf.

EIA. (2021). *Overview – Brazil*. Available at https://www.eia.gov/international/analysis/country/BRA.

EPE – Empresa de Planejamento Energético. (2020). *Analysis of Biofuels´ Current Outlook – Year 2019*. Ministério de Minas e Energia, EPE, Rio de Janeiro, July 2020, 80 p. Available at https://www.epe.gov.br/sites-pt/publicacoes-dados-abertos/publicacoes/PublicacoesArquivos/publicacao-489/Analysis_Biofuels_Current_Outlook_2019.pdf.

EPE – Empresa de Planejamento Energético. (2022). *Plano Decenal de Expansão de Energia-2031*. Ministério de Minas e Energia, EPE, Rio de Janeiro, 411 p.

EQUIPAV. (2009). *Uso da palha da cana-de-açúcar*. EQUIPAV, São Paulo.

GAZZAFFI, R.; OLIVEIRA, K. M.; SOUZA, A. P. de; GARCIA, A. A. F. (2010). Melhoramento Genético e Mapeamento da Cana-de-Açúcar. In CORTEZ, L. A. B. (coord.) (2010). *Sugarcane Bioethanol R&D for productivity and sustainability*, Blucher, 992 p.

GOLDEMBERG, J. (2008). *The Brazilian Biofuels Industry, Biotechnology for Biofuels 2008*, 1:6 doi:10.1186/1754-6834-1-6. Available at https://biotechnologyforbiofuels.biomedcentral.com/articles/10.1186/1754-6834-1-6.

LEITE, R. C. C.; LEAL, M. R. L. V.; CORTEZ, L. A. B.; GRIFFIN, W. M.; SCANDIFFIO, M. I. G. (2009). Can Brazil replace 5% of the 2025 gasoline world demand with ethanol? *Energy 34,* 655-661.

MACHADO, A. (2021). *Brasil produz milho em três safras por ano*. StoneX. Available at https://www.mercadosagricolas.com.br/inteligencia/brasil-produz-milho-em-tres-safras-por-ano/.

MME – Ministério de Minas e Energia. (2021). *Balanço Energético Nacional*, MME, EPE, Rio de Janeiro. Available at https://www.epe.gov.br/sites-pt/publicacoes-dados-abertos/publicacoes/PublicacoesArquivos/publicacao-601/topico-596/BEN2021.pdf.

MUTTON, M. A.; ROSSETTO, R.; MUTTON, M. J. R. (2010). Agricultural Use of Stillage. In CORTEZ, L. A. B. (coord.). *Sugarcane bioethanol — R&D for Productivity and Sustainability*, São Paulo, Blucher.

NIPE-UNICAMP; IBGE; CTC. (2017). *Mapa da Produção de Cana-de-Açúcar no Brasil*. Available at: https://www.researchgate.net/figure/Figura-1-Mapa-da-producao-de-cana-de-acucar-no-Brasil-Fonte-NIPE-Unicamp-IBGE-CTC_fig1_339347290.

NOVACANA. (2020). *Projeto de etanolduto da Logum Logística está disponível para comentários*. Novacana, Campinas/SP. Available at https://www.novacana.com/n/etanol/logistica/projeto-etanolduto-logum-logisticadisponivel-comentarios-090620.

ROSILLO-CALLE, F. FURTADO, P.; REZENDE, M. E. A.; HALL, D. O. (1996). *The Charcoal Dilemma*: Finding a sustainable solution for Brazilian industry, Intermediate Technology Publications, London.

SEEG – Sistema de Estimativas de Emissões e Remoções de Gases de Efeito Estufa. (2021). *Análise das emissões brasileiras de gases de efeito estufa e suas implicações para as metas climáticas do Brasil 1970 – 2020*, by POTENZA, R. F. et al, SEEG. Available at https://www.oc.eco.br/wp-content/uploads/2021/10/OC_03_relatorio_2021_FINAL.pdf.

UNICA. (2009). *Mapa da distribuição da cana-de-açúcar no Brasil*, UNICA. Available at https://www.researchgate.net/figure/Figura-21-Mapa-da-distribuicao-da-cana-de-acucar-no-Brasil-Fonte-UNICA-2009_fig1_331843202.

CHAPTER 4
FUTURE PRODUCTION MODELS OF BIOFUELS IN BRAZIL

4.1. INTRODUCTION

As seen in previous chapters, Brazil created sustainable biofuels production models, both for ethanol and biodiesel. However, these models need to be revised or adapted to allow further expansion of biofuels. This could be done either by modifying existing ones or the development of new market models.

Sugarcane ethanol is the most important biofuel in Brazil. Its economics were heavily linked to the sugar market. During the last few decades, the success of new ethanol-built distillery, was directly associated with joint production of sugar. This model currently has its limitations.

Also, criticism related to biofuel and food competition, although not necessarily applicable in the case of sugarcane ethanol, do raise some concerns. Brazil also faces important criticism related to land use, particularly with deforestation in the Amazon region and beef cattle ranches in pastureland. Therefore, there could be a way out by reconciling all these issues by introducing different production models.

4.2. NEW PRODUCTION MODELS FOR BIOFUELS

Bioenergy, specifically biofuels, were developed from existing feedstocks together with conversion technologies. This was the case for ethanol, biodiesel, and biogas and will probably be a good start for biokerosene and other fuels. Therefore, biofuels developed from adaptations, somehow inherited positive and negative features of their "mother" raw material and their "mother" conversion technology. In other words, biofuels were not designed from zero, in which environmental sustainability criteria were of little concern. Should biofuels have been conceived with environmental sustainability in mind, the outcome would have been different.

Perhaps we need rethink the entire production and conversion process as the current situation is very different. It could be said that the difficulties presented today may also represent a great opportunity for tomorrow. For example, sugarcane is a recognized as an excellent energy crop e.g., it produces, on average, 7,000 litres of ethanol/ha/yr besides the fibres/bagasse. But some studies have demonstrated a potential of 10,000 litres/ethanol/ha/yr, if G2 ethanol is coproduced (LEITE, 2009). What this example demonstrates is the considerable potential of sugarcane as a feedstock. With the development of new energy sugarcane varieties, this potential could increase even further, at least on the long term. Also, with the bagasse and leaves and tops, electricity can be produced in large scale. For example, by using combined cycles it is possible to produce at least 787.5 kWh/ton of cane (EQUIPAV, 2009). Besides ethanol and electricity, sugarcane has also shown the potential to produce other coproducts, such as animal feed (e.g., Vale do Rosário Mill), dried yeasts (São Martinho Mill), and biomethane (several mills are implementing new projects).

Other possibility is to extract the whole cane juice and instead of producing ethanol convert the juice directly to methane by a biodigestion process. Many experts consider vinasse biodigestion as a great contribution to the environment, while others consider biomethane as a precursor of green hydrogen. The remaining fibre (bagasse and trash) could then be used to generate heat and power.

All these are potential technological scenarios or pathways. Following is a more detailed discussion.

4.2.1 SECOND GENERATION ETHANOL OR G2 (USING BAGASSE AS RAW MATERIAL); CELLULOSE-BASED ETHANOL; COGENERATION; BIO-OIL PLANTS

Sugars, starches, and protein are noble raw materials. They should be used with parsimony since they are difficult to produce from a metabolic point of view. Plants are quite successful in producing sugar, starch, and protein to recreate life. Biofuels are just one clever way to convert these raw materials into useful energy. Human beings have been using firewood for long time, and this was, possibly, the first fundamental discovery we ever made. Firewood is still very much in use today (or “traditional biomass”), and in large quantities in many areas of the world, but predominantly in countries in Africa, Asia, and Latin America. Traditional biomass is basically extracted from biomes in quantities two or three times larger than how modern bioenergy is produced. Much of the traditional use of biomass is considered unsustainable and of very low energy inefficiency, and consequently should be phased out, but this is beyond the scope of this book.

To bypass the food vs biofuels dilemma, ligno-cellulosic materials are being seriously considered, particularly in the last two decades (see Chapter 1), to produce biofuels such as ethanol. Ligno-cellulosic can be converted into ethanol by two different ways: acid and enzymatic hydrolysis. There are other ways to produce biofuels using thermoconversion technologies, including the existing Rankine cycles, pyrolysis, and gasification. The experiences of ligno-cellulosic hydrolysis and thermoconversion in Brazil are discussed below.

4.2.1.1. Production of cellulosic or G2 ethanol

The production of G2 ethanol basically consists of two steps: the first is the hydrolysis f polysaccharides forming monosaccharides and disaccharides; secondly is the fermentation of produced sugars into ethanol. The difficulty is that the hydrolysis of cellulose, hemicellulose, and lignin generates products which ultimately may inhibit the fermentation process.[1] Therefore, an additional previous pre-treatment process is necessary to change or remove the lignin or the hemicellulose, increase its surface area, and decrease the cellulose polymerization degree and its crystallinity (Santos et al., 2012).

Probably, the biggest challenge in the G2 ethanol production process relates to pre-treatment. According to Santos et al an adequate pre-treatment process needs to have the following characteristics: a) low cost and investment; b) avoid the formation of undesired by-products which can inhibit the enzymatic action; c) allow high cellulose digestibility by the enzymes; and d) high recovery of sugars (SANTOS et al., 2012)

4.2.1.1.1. Production of ethanol using acid hydrolysis

The initial research works in acid hydrolysis started at CODETEC/UNICAMP at the end of 1970s with José Carlos Campana Gerez. This project reached pilot plant scale. And in the 1970s, COALBRA company[2] was created with the objective to tropicalize the soviet acid hydrolysis technology to produce ethanol from wood and animal protein as by-product. This demonstration plant achieved 30,000 litres of ethanol/day, located at the "*triângulo mineiro*" in the state of Minas Gerais. The COALBRA plant had technical problems such as the difficulty to dehumidify the eucalyptus lignin using filter press. In addition, the heavy investments in equipment resulted in high production costs (CORTEZ, 2016).

In 2001 DEDINI started the development of the DEDINI Rapid Hydrolysis Process (HDR) with the support of a FAPESP project.[3] The initial configuration was to use the acid catalysed Organosolv saccharification, integrating the G2 and the G1 ethanol. Among the difficulties was the problems posed by bagasse feeding technology.

4.2.1.1.2. Production of ethanol using enzymatic hydrolysis

As mentioned before, the FvF debates reborn the concept of producing sustainable biofuels from ligno-cellulosic raw material to avoid food competition. Ligno-cellulosic

---

1 https://www.scielo.br/j/qn/a/9n4nqyhZ3dVtZrpHQy5thDh/?lang=pt

2 The project was coordinated by Sérgio Motta, José Goldemberg, Gil Serra, and José Roberto Moreira as collaborators

3 https://fapesp.br/3257/bioen-chamada-fapesp-dedini-para-apoio-a-pesquisa

is by far more abundant than sugars, starches, or oils and are easier materials to convert to biofuels.

In respond to the need to develop hydrolysis, several laboratories initiated research into this area. This was particularly the case of the National Research Energy Laboratory (NREL), USA, focusing the enzymatic hydrolysis. Acid hydrolysis was considered a more difficult route for the conversion of biomass into ethanol or methanol. The NREL initiative influenced Brazilian researchers, particularly at the Instituto de Pesquisas Tecnológicas (IPT), University of São Paulo (USP), and UNICAMP. A research group- - Bioethanol Network- was put together under the leadership of Rogério Cerqueira Leite. This group worked to identify the level of competence, scientific and technology barriers to cellulosic ethanol in Brazil. (CORTEZ, 2016)

Another research group, also coordinated by Cerqueira Leite, was trying to answer the question of how much sugarcane ethanol could be produced in Brazil. A group of agronomists studied soils and climate conditions in different regions, excluding ecological sanctuaries such as the Amazon, Pantanal, and the remaining Atlantic Forest. A total of 17 areas (Figure 4.1) were selected as a function of the potential shown in the maps (Figure 4.2). These maps were plotted by agronomists from the Sugarcane Research Centre – CTC from Piracicaba, SP, based on local edaphoclimatic conditions.

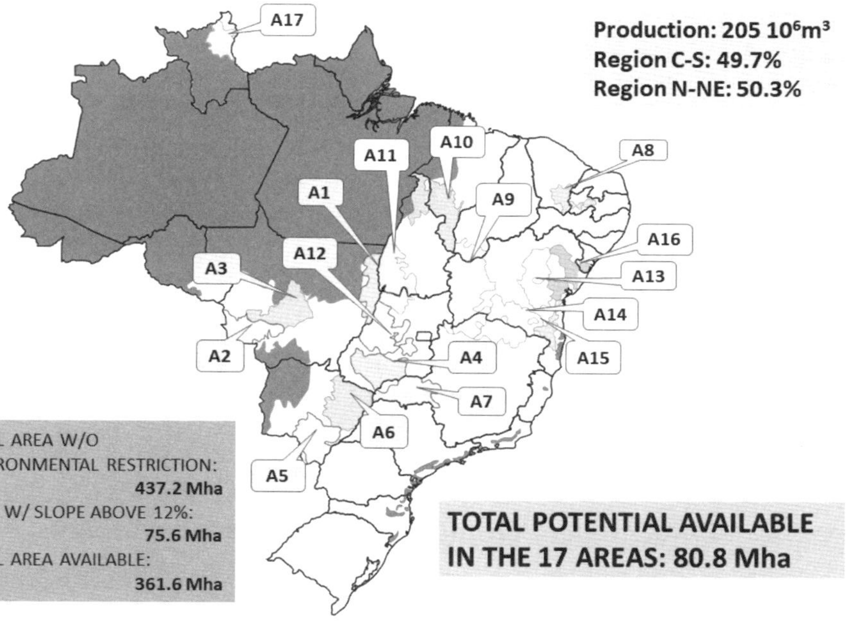

Figure 4.1 Seventeen selected areas for potential production of sugarcane in Brazil.

Source: Leite (2009).

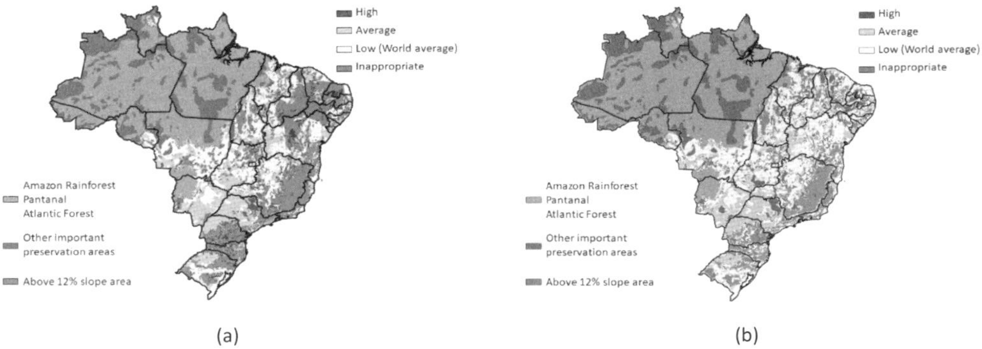

Figure 4.2 Potential for sugarcane production in Brazil, considering soils and climates, (a) non-irrigated, and (b) with survival irrigation. Source: Leite (2009).

Then, based on these 17 selected areas and average combined sugarcane ethanol productivity, and plantation areas were then estimated. The conventional technology, based exclusively on the fermentation of sugars, were 6,000 litres/ha in 2005 and expected to go up to 8,200 litres/ha in 2015. Regarding the hydrolysis technology, it was expected to produce 1,100 litres/ha in 2015 and 3,500 litres/ha in 2025. Unfortunately, for several reasons, these yields were never achieved, although G2 ethanol is progressing well in Brazil, as discussed below.

Naturally, if hydrolysis technology was to be successfully developed in Brazil, the required sugarcane area would have been significantly smaller. For example, to produce 205 billion litres of ethanol/year using only the existing 1G ethanol technology with 7,000 litres/ha/yr, approximately 29 Mha would have been necessary. However, if G2 ethanol could contribute with an additional 3,000 litres/ha/yr, totalling an overall yield of 10,000 litres of ethanol/ha/yr, the required area would be 20.5 Mha. So, it was clear at that time that a national R&DD effort was necessary to develop hydrolysis technology.

As a result of these discussions, the Brazilian Centre for Research in Energy and Materials (CNPEM), decided to create the Brazilian Bioethanol Science and Technology Laboratory (CTBE) funded by the federal government. The new centre, the CTBE, would have to contribute to assure the leadership in sustainable ethanol production and implied basically two axes: a) redesigning the existing agriculture, particularly helping to disseminate the existing "best practices", and helping to promote adequate sugarcane mechanization; and b) helping in the developing the G2 ethanol technology. Several partnerships were signed up with both agricultural and industrial sectors. Another interesting contribution included a) a pilot plant which allows the development of experiments with the objective of reproduce industrial conditions; and b) a virtual biorefinery, a simulation platform useful to evaluate different technologies in the sugarcane bioethanol sector. Today, the CTBE operates under the name of Brazilian Biorenewables National Laboratory (LNBR).

To boost G2 ethanol the Brazilian Development Bank (BNDES) launched in 2011, together with the Brazilian Innovation Agency (FINEP), a program to support industrial technology innovation in the sugar energy and sugar chemical sectors, the PAISS.[4] One of the 35 selected business plans financed by the PAISS program, was the GranBio project to produce G2 ethanol.

In 2014 the GranBio project (60 million litres/year capacity) started operations at São Miguel dos Campos, AL. The GranBio project uses "energycane" as feedstock, which presents a higher overall productivity with a fibre/sugar ratio of 3, as compared with conventional sugarcane varieties with 2. With higher fibre content it could make sense to develop a process such as hydrolysis. The expected overall yield for the Granbio project was 10,000 litres/ton of sugarcane: nearly 45% higher than conventional sugarcane. This project received technology support from partners such as Beta Renewables, Novozymes, and DSM.

Figure 4.3 GranBio G2 Ethanol plant in São Miguel dos Campos, AL. Source: GranBio.[5]

RAÍZEN ENERGIA was founded in 2011 by integrating part of the COSAN (50%), at that time the largest producer of sugar and ethanol in Brazil, with the Brazilian SHELL (50%). RAÍZEN ENERGIA amalgamated the COSAN units, responsible for sugar, ethanol, and cogeneration, together with the distribution and commercialization operations of fuels of both companies. RAÍZEN presently is the largest fuel integrated group in Brazil, responsible from sugarcane production to the fuel distribution. The group had 35 sugar mills processing 105 million tons of cane in 2019/2020 harvest.[6]

The RAÍZEN group started its industrial production of G2 ethanol in 2015. An annexed G2 distillery was installed alongside the Costa Pinto Mill in Piracicaba, SP (Figure 4.4). The main idea with the annex plant is to have synergy between the G1 and

4 https://www.bndes.gov.br/wps/portal/site/home/imprensa/noticias/conteudo/20130507_etanol

5 http://www.granbio.com.br/conteudos/bioflex-biocombustiveis/

6 https://conteudos.xpi.com.br/renda-fixa/cra-raizen-dez-2022/

G2 processes to save energy. Its production capacity is 40 million litres/year. According to the RAÍZEN group they are presently the largest global cellulosic ethanol producer and trader in a commercial scale. Just recently the RAÍZEN group signed a contract to sell 460 million litres of G2 ethanol in the next 9 years. The group has, up to now, commercialized nearly 1 billion litres of G2 ethanol.[7]

Both GranBio and RAÍZEN have faced technical difficulties mainly related to pre-treatment and lignin filtering. Although significant improvements have been achieved, both units are still operating under capacity (EPE, 2022).

Figure 4.4 RAÍZEN G2 ethanol plant in Piracicaba, SP. Source: G1 (2015).

The RAÍZEN group recently announced the construction of a second G2 ethanol plant together with the Bonfim mill at Guariba, SP. The new plant will have the capacity to produce 82 million litres of ethanol/harvest, nearly double of the first plant.[8] The G2 ethanol business is progressing well, as RAÍZEN group declares that their prices receive a premium of up to 70% in the North American market.[9] Under such conditions, the same source confirms G2 ethanol competitiveness. These high premiums are related to the renewable identification numbers (RIN) associated with the technology under the RFS program and how EPA tracks and evaluates the entire process.[10] Table below shows a comparison of RAÍZEN´s ethanol emissions reduction and fuel ethanol produced in different regions and different feedstocks.

7 https://www.novacana.com/n/etanol/2-geracao-celulose/raizen-anuncia-venda-460-milhoes--litros-etanol-celulosico-100821#:~:text=Atualmente%2C%20a%20Ra%C3%ADzen%20possui%20uma,milh%C3%B5es%20de%20litros%20por%20ano.

8 https://www.udop.com.br/noticia/2021/06/25/raizen-vai-construir-sua-segunda-planta-de-etanol--celulosico.html

9 https://udop.com.br/noticia/2021/10/26/raizen-esta-vendendo-etanol-celulosico-com-premios--de-70.html

10 https://www.epa.gov/fuels-registration-reporting-and-compliance-help/rin-trades-and-price-information

Table 4.1 Carbon footprint, reduction of emissions, and volume of ethanol produced in different regions. Source: (RAÍZEN, 2022).

G2 Ethanol	Raízen (ACV Br/KPMG)[40]	13.6	86%	Raízen 20 billion liters[10]
	Watanabe et al. 2015 [46]	12.9	85%	
Sugarcane Ethanol	Raízen (ACV Br/KPMG)[40)	19.6-20.9	78-79%	34 billion liters in Brazil and Raízen 2.5 billion liters [47]
	Pereira et al. 2019 [39]	16-24	74-83%	
Sugarcane and corn ethanol	Milanez et al. [43]	28-31.4	67-70%	Distributed between sugarcane and corn production*
Brazilian corn ethanol	Moreira et al. (2020)	18.3-25.9	72-81%	Brazil 1.7 billion liters [47]
US corn ethanol	Pereira et al. 2019 [39]	43.4-61.9	34-54%	USA 59.7 billion liters [48]
European corn ethanol	Hamelinck et al. 2019 [41]	42.5	55%	Europe 2.7 billion liters [49]
	RED II, 2018 [35]	31.4-71.7	24-67%	
European sugar beet ethanol	Manochio et al., 2017 [42]	40	57%	Europe 1.08 billion liters [49]
	RED II, 2018 [35]	34-50	47-64%	
Gasoline	ANP	85.4		
	RED II, 2018 [35]	94		

Key: The table above shows the carbon footprint of ethanol in different regions (Brazil, USA, and Europe) from different inputs (sugarcane, second-crop corn, US and European corn, wheat, and sugar beet) compared to gasoline. Some figures have been selected for methodological consistency and to facilitate comparisons, presenting carbon footprints with energy and economic allocations. Data certified by RenovaBio were not reported because the mills opted for different strategies. The pursuit of higher volumes of CBIOs can lead to the inclusion of producers with lower quality data, which increases the carbon footprint. *The volumes of ethanol produced by the "flex mills" are distributed between sugarcane and corn production.

4.2.1.2. Combustion of ligno-cellulosic materials (existing Rankine/sugarcane mills)

Sugar and ethanol mills are the largest installed facilities to produce electricity from biomass in Brazil. According to the Brazilian Ministry of Mines and Energy annual report (MME, 2021), biomass represented 9.1% of all electricity used in the country in 2020.[11] In 2021, sugarcane bioelectricity provided 79% of all biomass electricity (RE-hydro with 65.2%, wind with 8.8%, and solar with 1.7%) representing 84.8% of all electricity.

As mentioned above, the challenge refers to how the energy matrix will behave with the generation of electricity from renewables. The situation with hydroelectricity is

11 https://summitagro.estadao.com.br/noticias-do-campo/cana-de-acucar-gera-energia-eletrica-no-brasil/#:~:text=A%20gera%C3%A7%C3%A3o%20de%20energia%20pelo,biomassas%20com%204%2C1%25.

complicated because, although there is still significant potential to be explored, it is concentrated in the Amazon region where the topography imposes large dams and large flooded areas, with significant impacts on biodiversity and the environment. The possibilities are much better with wind, particularly in the Northeast region where there is still a significant potential to be explored, and solar, a market that is growing fast, though still making a modest contribution. Biomass-based electricity has long tradition but depends very much on how the sugar and ethanol market will behave/develops. Presently biomass electricity is growing a much slower pace than wind and solar.

Two important points to remember is that the South-East and Central-West regions suffer from chronic droughts, particularly so in the last 20 years due to increasing population and climate change. The existing water reservoirs are used both for water supply and electricity generation which results in conflicting objectives. Therefore, biomass-based electricity can play an important role, particularly because sugarcane bagasse, the main feedstock, is harvested in the drought months, typically from May to October. Sugarcane itself, although a relatively resilient crop, also suffers from the consequences of severe and prolong droughts.

In 2002-2003, during a major drought, a study was conducted to evaluate the potential contribution of sugarcane to the production of electricity, concluding that the sugar and ethanol sector can significantly contribute to bioelectricity production. For example, in 2021, only 60% of all existing 369 mills were exporting electricity to the national Grid. If all mills sold electricity to the grid, the total amount could reach 151 GWh, meaning 26% of all electricity consumed in Brazil.[12] These figures were calculated by the Brazilian Energy Research Office (EPE), published in the "Bioeletricidade em Números" report from February 2022 by the Observatório da Cana (UNICA, 2022). One the difficulties to the expansion of cogeneration using sugarcane bagasse has been the low price paid for MW, which has been based on average electricity cost in Brazil.

4.2.1.3. Other thermoconversion technologies (pyrolysis and gasification)

The pyrolysis and gasification processes are thermoconversion technologies which can transform biomass into valued-added products. They have been tested in Brazil in the last decades and are now close to commercial utilization.

Sugarcane- biomass, consists basically of the fibrous material of bagasse and trash (cane tops and leaves). Figure 4.5 shows how a thermochemical process can be integrated to an existing sugar & ethanol mill.

12 https://summitagro.estadao.com.br/noticias-do-campo/cana-de-acucar-gera-energia-eletrica-no--brasil/#:~:text=A%20gera%C3%A7%C3%A3o%20de%20energia%20pelo,biomassas%20com%204%2C1%25.

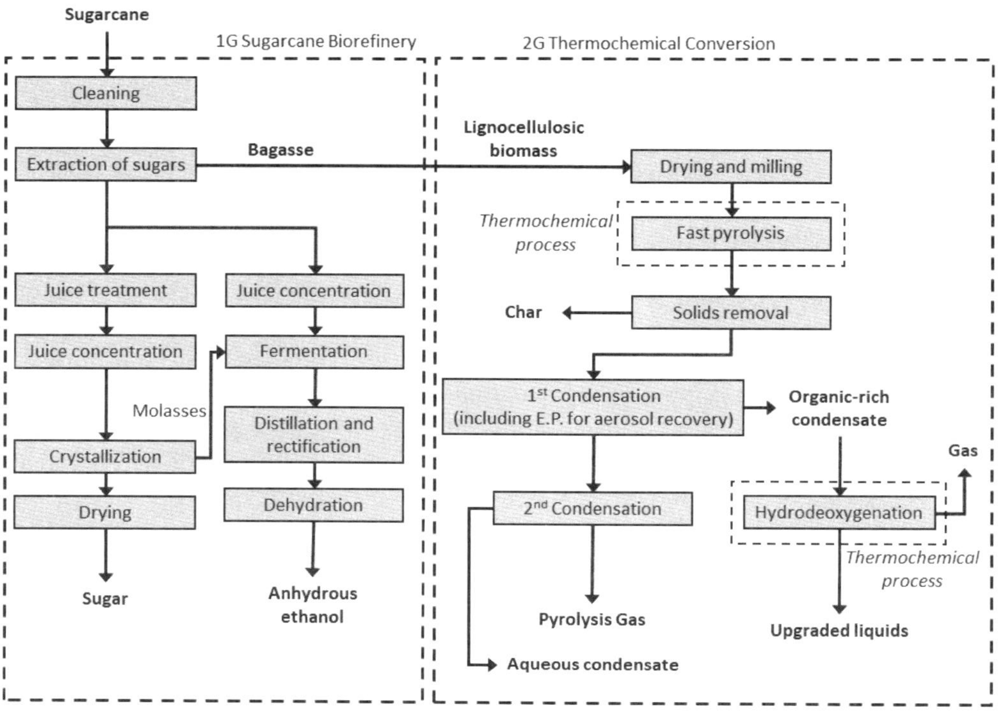

Figure 4.5 Integration of a thermochemical process with an existing sugar mill.
Source: KIT and IPT (2022).

At the end of 1970s, under the initiative of CESP and coordinated by of Eduardo Sabino de Oliveira, a wood electric gasifier was built to produce syngas for conversion to methane. The plant was installed in Corumbataí, SP, in 1987, under the leadership of José Roberto Moreira. The plant consisted of several large size gasifiers (capacity 20 tons of wood/h), but it was discontinued (CORTEZ, 2016), perhaps due to operational problems and economic viability issues.

The Institute for Technology Research (IPT) has made several contributions, both to G2 ethanol and thermoconversion processes.[13] One of the most important was a Biomass Integrated Project (PIB) coordinated by Vicente Mazzafera. This project investigated elephant grass as an alternative feedstock for bioenergy.[14] The research groups focused basically on agronomic and energy conversion studies. The prospects were positive, although no commercial application was actually made.

Also, in the early 1980s, the Brazilian company TERMOQUIP started operation in Campinas, SP, based on know-how developed at UNICAMP by Saul D´Ávila and

13 http://webcache.googleusercontent.com/search?q=cache:yJDWixEQqoAJ:www.ipt.br/download.php%3Ffilename%3D697-Biomassa_para_producao_de_energia_Colloquium_SAE_Brasil__Energia_Verde.pdf+&cd=26&hl=pt-BR&ct=clnk&gl=br

14 https://sites.google.com/a/capimelefante.org/www/trabalhos-apresentados/t5

Themístocles Rocha. TERMOQUIP manufactured biomass gasifiers for clients including PETROBRAS. Also, fast pyrolysis rectors were made for experimental work using sugarcane bagasse and whole sugarcane as feedstocks (CORTEZ, 2016).

In the last decade, an important project has been conceived by the Canadian ENSYN with SUZANO (Fíbria). Production was planned to reach 110 million litres of biocrude/day.[15]

The main idea with this technology is to investigate direct conversion of biomass into a useful biofuel, not necessarily ethanol or biodiesel. If this technology works, it could provide a boost to biomass-based fuels given the large amount of bagasse and tops and leaves from sugarcane alone potentially available, not only in Brazil but worldwide. Potential candidates include "biocrude and bio-oil". This could also have a major economic impact in existing refining process.

Just for illustration, if this viable biomass such as sugarcane, eucalyptus, or other energy crops could eventually be converted into biocrude and then enter in an oil refinery to produce the same oil derivates as gasoline, diesel, and kerosene, this will have a significant impact, just by partially using this new source.

This model could be used by PETROBRAS if it is to become, gradually, a green oil company, since the company has excellent conditions and experience to use ethanol and sugarcane as tools to innovate in this area.

Also, as mentioned in Chapter 1, Brazil used to have a small, but flourishing ethanol-based chemical sector, particularly fine chemicals. Presently, large investments are directed into biomass-based feedstocks, and away from fossil fuels-based ones. An example is the well-succeeded Braskem polyethylene (PE), polypropylene (PP) and polyvinyl chloride (PVC) resins. This represents a great opportunity, particularly for ethanol and its use for new green materials. See BNDES (2008) for further details on ethanol-chemical industry in Brazil.

4.3. THE EXPANSION OF CORN ETHANOL IN BRAZIL

4.3.1. LAND USE, DEFORESTATION, AND GHG EMISSIONS FROM BEEF CATTLE IN BRAZIL

Land use in Brazil is dominated by native forests (65%) and pastureland (18%).[16] These numbers are subject to some variations e.g., pastureland maybe closer 20% of the Brazilian territory, approximately 170 Mha. There is a significant controversy regarding the size of Brazilian beef cattle and the area it occupies. The 170 Mha of pastureland is, in fact, a conservative figure. The exact number is difficult to know, but it is probably anything from 150 to 200 Mha (Figure 4.6).

15 https://www.gazetaonline.com.br/noticias/economia/2017/12/es-tem-99-9-de-chance-de-receber--fabrica-de-biopetroleo-diz-fibria-1014110923.html

16 https://mapbiomas.org/infograficos-1?cama_set_language=pt-BR

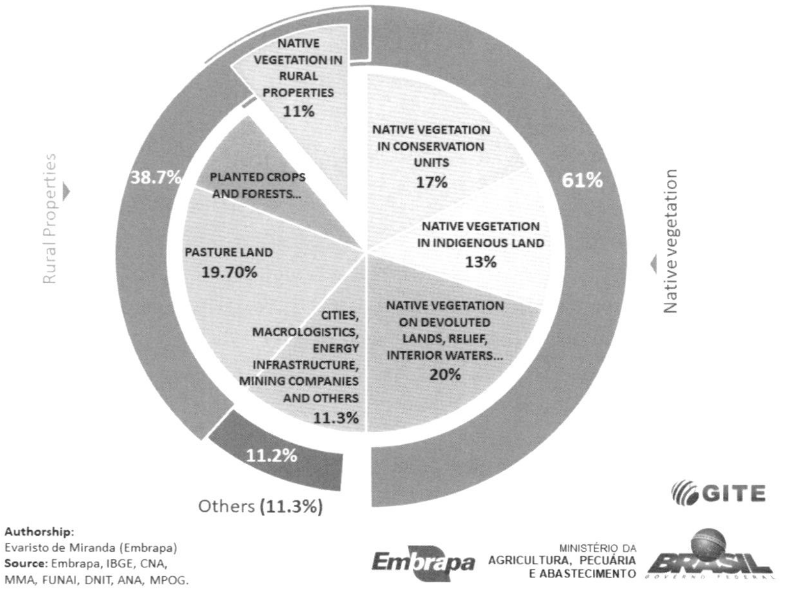

Figure 4.6 Land occupation in Brazil. Source: EMBRAPA.[17]

Another important issue regarding pastureland in the Amazon region refers to deforestation and its core causes. Several sources indicate that beef cattle expansion happened mainly in North and Central-West regions and are considered one of the key causes of Amazon Forest deforestation (IPAM, 2016).

As for beef cattle herd size, according to IBGE, MAPA, USDA, FAO, and Athenagro, estimates, this varies from 190 to 250 million heads, making Brazil the first or second largest herd in the world. Pastureland and beef cattle ranches are intrinsically and environmentally related, as land use is correlated with GHG emissions. This large heard represents nearly 17% of Brazilian GHG emissions. As discussed in the previous chapter, the Brazilian agriculture sector is the second most important source of GHG emissions, with 27% of total, the first being land use change. For clarity, the 27% discussed in the previous chapter, includes GHG emissions of both agriculture (10%) and beef cattle (17%).

For the Brazilian beef cattle industry, the need to reduce its GHG emissions also creates opportunities to intensify and improve the production systems, greatly benefiting the meat market. One of the largest beef cattle producers, JBS, has launched a sustainability strategy which allowed to recycle 25 thousand tons of solid residues (mainly tallow), and an additional 2 billion litres of biodiesel, plus fertilizers and organic products (Folha de São Paulo [18]). This affirmative action could be partly an answer to criticism, both domestic and international.

17 https://www.beefpoint.com.br/no-brasil-a-salvacao-da-biodiversidade-do-meio-ambiente-e-da-economia-esta-na-agropecuaria-diz-pesquisador-da-embrapa/

18 Folha de São Paulo, "Terra", JBS advertisement, April 22th, 2022, Estúdio Folha Terra, p 4-5.

Beef cattle GHG emissions are mainly methane originated from the animal's enteric fermentation[19] and nitrous oxide originated from the soil (Figure 4.7 below). The nitrous oxide may come from the application of nitrogen as fertilizer and from beef cattle manure.

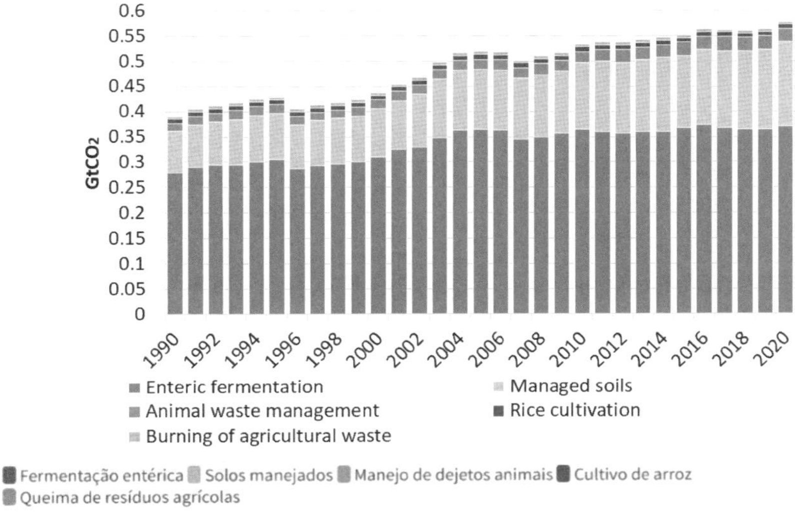

Figure 4.7 Evolution of GHG emissions by the Brazilian beef cattle, in Gtons of $CO_{2equivalent}$ (1990-2020). Source: SEEG/Observatório do Clima (2022).

The reduction of Brazilian GHG emissions is therefore very much related to improvements of the beef cattle industry. This improvement and modernization need to be accelerated, so that the country can increase beef cattle productivity and recover degraded pastureland areas. Another important measure is through genetic improvement to allow more intensive, and confined production systems and reduction of life cycle (slaughter age) and pastureland management. Naturally, it is important to make aware to cattle producers, with their particularities e.g., such as size, or production systems, of the importance of sustainable production while at the same be economically viable.

4.3.2. CORN ETHANOL IN BRAZIL

Until recently, corn ethanol did not enjoy a good reputation in Brazil. This was probably due to the relatively bad energy balance of corn ethanol[20] when compared with sug-

19 Brazil is among the 100 countries which in the Conference of Climate – COP26 last Nov 2021 Glasgow, Scotland, agreed to reduce methane gases emissions in 30% until 2030. Last March 2022, the Brazilian Federal government launched a national program (Methane Zero) to reduce methane originated from organic residues.

20 According to RFA report, the corn ethanol energy balance is improving, basically due to more accurate estimations on the energy used in the process, particularly regarding the distillers' grains.

arcane. The main reason for this negative energy balance is based in the fact that corn, when used as feedstock, requires external use of energy to run the mill. Sugarcane, on the contrary, has abundant bagasse and fibre (representing nearly 2/3 of total energy in the plant). If energycane varieties are used, this proportion can go up to ¾. In addition, if boilers are run with maximum efficiency, this could generate large energy surpluses.

Recently, in some areas located in the country agricultural frontier (Mato Grosso and Goiás states) are more interested in ethanol production systems. This is basically due to low corn prices in that region, which at the same time are due to logistic difficulties associated with transporting corn grains to the South-East region where there is greater market demand.

Besides, in several states in Central-West and North, ethanol (E100) prices are higher making practically inviable the use of flex-fuel engines using E100. Gasoline (E27.5) are also high, but when the consumer makes his calculations,[21] the decision favour E27.5 (Figure 4.8). Therefore, it seems logical to explore the existing gap (low corn prices and high ethanol prices) and to produce corn ethanol.

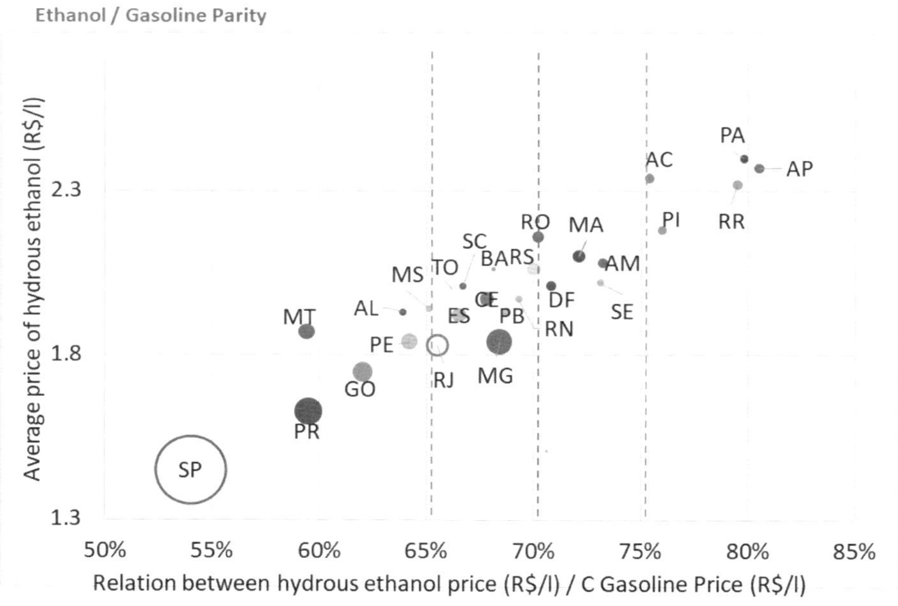

Note: The bubble sizes are proportional to the ethanol consumption from July 2001 to December 2009. Prices are expressed in Reais of December 2009. The dashed red line (70%) represents the ethanol-gasoline parity, and the dashed green lines shows the 65% and 75% band.

Figure 4.8 Average price of E100 and gasoline-ethanol price parity for different states in Brazil – 2001 to 2009. Source: Souza (2010) using ANP data.

21 The rule of thumb: when ethanol E100 prices are higher than 70% of gasoline E27.5 prices, it is worth to use gasoline E27.5.

One important point with regard to production and use of sugarcane ethanol is that sugarcane is not a countrywide crop. In fact, sugarcane plantations are almost 80% in Central-South region (mainly in São Paulo, Paraná, Minas Gerais and Goiás states) and around 20% in the Northeast (mainly in Alagoas, Pernambuco and Paraíba states). This geographic distribution influence ethanol prices because of higher transportation costs. Typically fuels costs are nearly 50%, or even higher, compared to state of São Paulo, the centre of ethanol production and consumption and where most refineries are located. This, naturally, affects energy security and logistics since fuel needs to travel thousands of kilometres to destination. Also, often ethanol is imported from USA to supply the Amazon region. On the other hand, corn can be planted in more states and could complement sugarcane as feedstock for ethanol production (Figure 4.9 and Table 4.2 below).

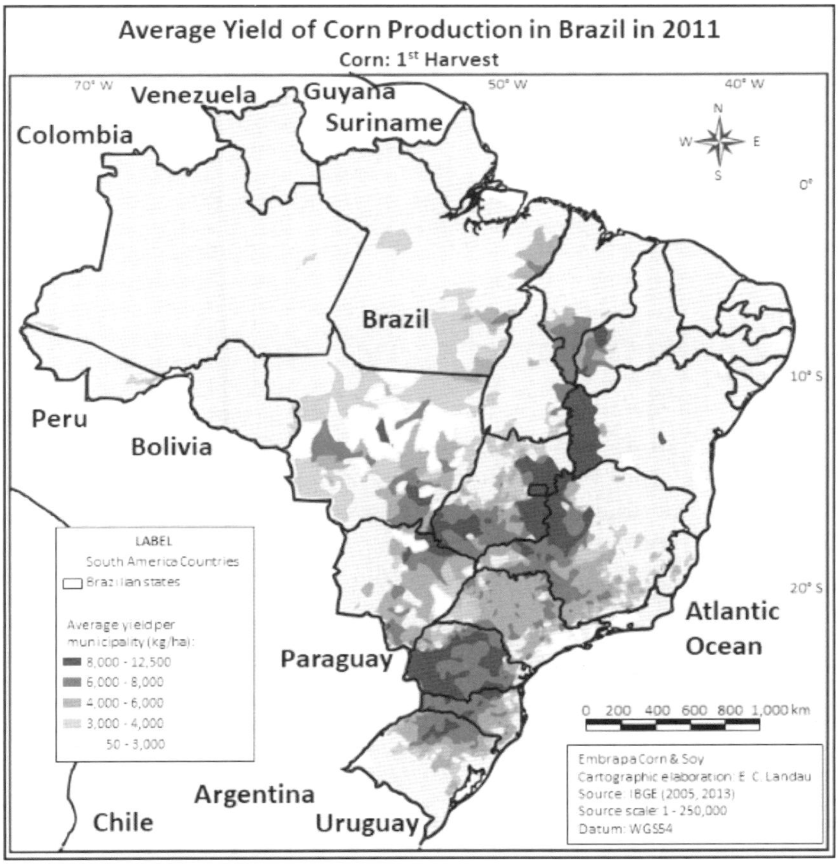

Figure 4.9 Average yield of Corn in Brazil in 2011, 1st harvest. Source: EMBRAPA[22]

22 http://panorama.cnpms.embrapa.br/mapas/produtividade-do-milho/mapa_milho1safra_rendimmedio2011_.jpg/view

Table 4.2 Corn Production in Brazil. 1st and 2nd crops. Source: CONAB (2022)

Year	First Crop	Second Crop
2000/01	35.8	6.5
2001/02	29.1	6.2
2002/03	34.6	12.8
2003/04	31.6	10.6
2004/05	27.3	7.7
2005/06	31.8	10.7
2006/07	36.0	14.8
2007/08	40.0	18.7
2008/09	33.7	17.3
2009/10	34.1	21.9
2010/11	34.9	22.5
2011/12	33.9	39.1
2012/13	34.6	46.9
2013/14	31.7	48.3
2014/15	30.3	64.0

One important feature of corn ethanol is the production of solid high-protein residues. There are two basic processes to produce corn ethanol in the USA, wet and dry. In the wet process the result is wet distillers' grains (WDG) and in the dry is distillers' grains (DDG). Typically, 1 ton of corn yields 425 litres of ethanol and 312 kg/ of DDG.[23] In principle the same yields are expected in Brazil.

Presently, several mills are already in operation in the country. Most of them are in the states of Mato Grosso and Goiás (Figure 4.10). In addition, Table 4.3 provides a list of all corn ethanol mills, either processing corn only or corn and sugarcane (flex mills).

23 https://sistemafaeg.com.br/faeg/noticias/etanol-de-milho/goias-deve-produzir-392-milhoes-de--litros-de-etanol-de-milho

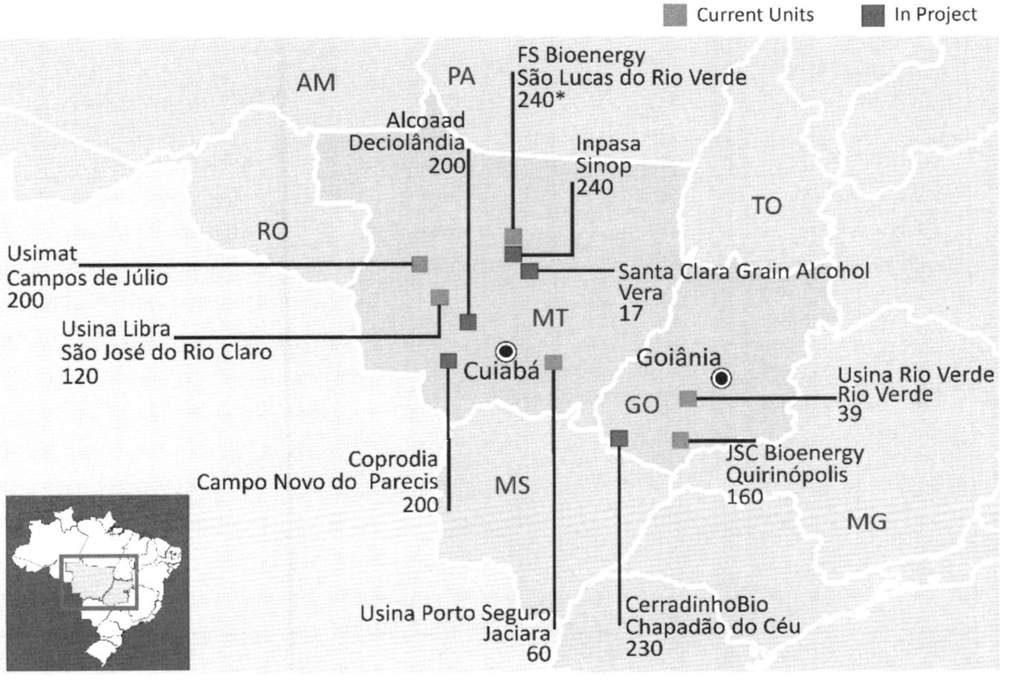

Source: Companies, Unem, Sindalcool-MT. *With expansion, capacity goes to 530 million litres per harvest.

Figure 4.10 Localization and production capacity of few corns' ethanol plants in Mato Grosso and Goiás.

Table 4.3 Main corn ethanol distilleries in the states of Mato Grosso and Goiás, Brazil.
Source: Compiled by the authors.

	Localization	feedstock	Production capacity of ethanol and DDG
USIMAT	Campos de Júlio, MT	Corn/Cane	480-520 thousand litres/day + DDG
Grupo Safras	Sorriso, MT	Corn	13.86 million litres of ethanol/year
Destilaria Libra	Diamantino/São José do Rio Claro	Corn/Cane	600 thousand litres/day
Fiagril/Summit FS Bioenergia	Lucas do Rio Verde, MT	Corn	210 thousand litres ethanol, 180 thousand tons of feed, 6 thousand tons oil/day
Usina Pantanal/Porto Seguro	Jaciara, MT	Corn/Cane	16,9 thousand litres/day+9.102 tons of DDG
Santa Clara Álcool de Cereais	Vera, MT	Corn/Sorghum/Rice	50 thousand litres/day
Ethanol S.A. Bioenergia	Nova Mutum, MT	Corn	200 million litres/year
ETAMIL (Coprodia Group)	Campo Novo do Parecis, MT	Corn/Cane	300 thousand litres/day

<table>
<tr><th colspan="2"></th><th>Localization</th><th>feedstock</th><th>Production capacity of ethanol and DDG</th></tr>
<tr><td colspan="2">Cooperativa Agroindustrial Deciolândia (Alcooad)</td><td>Tangará da Serra, MT</td><td>Corn</td><td>200 million litres/year</td></tr>
<tr><td colspan="2">INPASA</td><td>Sinop, MT</td><td>Corn</td><td>6 million litres/day</td></tr>
<tr><td colspan="2">VMG Bioenergia</td><td>Jataí, GO</td><td>Corn</td><td>7 million litres/month</td></tr>
<tr><td colspan="2">GEM Alimentos</td><td>Acreúna, GO</td><td>Corn</td><td>140 million litres/year</td></tr>
<tr><td colspan="2">Grupo São Martinho</td><td>Quirinópolis, GO</td><td>Corn/Cane</td><td>200 million litres/year</td></tr>
<tr><td colspan="2">CerradinhoBio</td><td>Chapadão do Céu, GO</td><td>Corn</td><td>230 million litres/year+150 thousand tons of DDG</td></tr>
<tr><td rowspan="2">São João-Cargill (SJC Bioenergia)</td><td>Usina São Francisco</td><td>Quirinópolis, GO</td><td>Corn/Cane</td><td rowspan="2">500 million litres/year + FlexyPro, FlexyDDG, FlexyDDG Plus</td></tr>
<tr><td>Usina Rio Dourado</td><td>Cachoeira Dourada, GO</td><td>Corn/Cane</td></tr>
<tr><td colspan="2">Santa Helena</td><td>Santa Helena de Goiás, GO</td><td>Corn</td><td>400 thousand litres/day</td></tr>
<tr><td colspan="2">Caçú</td><td>Vicentinópolis, GO</td><td>Corn/Cane</td><td>500 thousand litres/day</td></tr>
<tr><td colspan="2">Usina Rio Verde (Decal)</td><td>Rio Verde, GO</td><td>Corn</td><td>55 million litres/year</td></tr>
</table>

Production of corn ethanol reached 262.3 million litres in Goiás and 1.98 billion litres in Mato Grosso in 2021/22 season, about 2.2 billion million litres in both states.

Corn ethanol offers several advantages. According to Ricardo TOMCZYK, Director of the União Nacional do Etanol do Milho (UNEM), one of the greatest advantages of corn ethanol is that the production capacity can be quickly scale up e.g., it can be put into production in 1.5 years while sugarcane needs 4 years to implement a new plantation (NOVACANA, 2019).[24] According to the same source corn ethanol production can reach 8 billion litres in 2028 in Brazil.

4.3.3. POSSIBLE LAND USE CHANGE ASSOCIATED WITH CORN ETHANOL INTEGRATION WITH BEEF CATTLE IN BRAZIL

The most important issue concerning corn ethanol production are the possible impacts on land use. Such impacts will depend on how this integration takes place, or how much land will be spared e.g., the greater the animal density the greater land

24 https://www.novacana.com/n/eventos/ricardo-tomczyk-unem-desafios-etanol-milho-mato-grosso--brasil-030719

availability.[25] In addition, the correct localization of new distilleries will also affect the economics of this integration

What is necessary is to investigate the potential of the final indirect land use change (iLUC). Searchnger et al. (2008) introduced the land use change and the indirect land use concepts. According to the author, when a biomass is planted it will ultimately cause two effects: a LUC and an iLUC.

The LUC can be defined as the change in land use when another crop is grown. It is the direct effect; one crop substitutes the other crop. For example, LUC happens when a given crop, such as biomass "pushes" a previous crop or pasture. This is the "direct" land use change.

The iLUC refers to the second effect, the "land whose ultimate purpose is essentially changed from its previous use".[26] An "indirect" land use change can be and understood as a kind of "domino effect". According to the same source: "an example would be a forest land that was cleared for the cultivation of biofuel crops". Figure 4.11 illustrate the LUC and the iLUC.

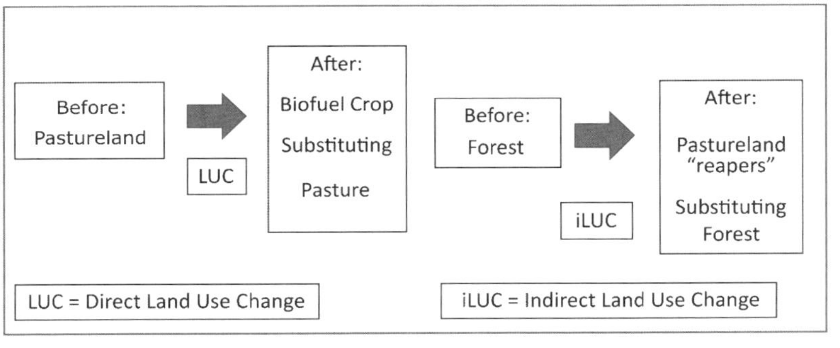

Figure 4.11 Schematic view of LUC and iLUC processes. Source: by the authors.

Considering that corn ethanol integration with beef cattle will shrink pastureland and free up land, there will be a "negative iLUC". In this case, the mitigated GHG emissions should be added and not subtracted. UC impact on sugarcane LCA results in mitigation to drop from 80-90% to about 60%. This is still in makes sugarcane ethanol as "advanced fuel" according to EPA.

What it is being proposed is a reverse strategy in which corn integration with beef cattle can result on the opposite effect, freeing up land, rather than causing deforestation. Naturally, this will benefit corn ethanol new LCA, improving the potential benefits when compared to the LCA corn ethanol produced in the USA.

To demonstrate this possible benefit, it is illustrated on Table 4.4 by the following example, based on a 335 M litres/year distillery installation. There are two options:

25 Of course, we are aware that when animal density is large, this also presents problems. Bit this is not necessarily the case in Brazil,

26 https://farm-energy.extension.org/what-is-direct-land-use-or-direct-land-use-change/

100% sugarcane or 100% corn as feedstock. Some assumptions are made to ensure a rough estimation of the installation of this fictitious ethanol plant. The investor is keen to know which option will be best e.g., LUC and expected iLUC.

Table 4.4 Comparison on sugarcane and corn ethanol indirect land use (iLUC) for a 335 M litres/day ethanol distillery. Source: by the authors.

	Sugarcane	**Corn**
Planted area for the raw material	50 thousand ha + reform area	140 thousand ha
Planted area for extra fuel (eucalyptus)	No area needed since sugarcane produces abundant fibre (bagasse)	25 thousand ha
Total planted area	67 thousand ha	165 thousand ha
Impact on land use	67 thousand ha will need to be open to cultivate sugarcane	214 thousand ha of pastureland will chance use (LUC)
Balance of areas (iLUC)	(–) 67 thousand ha of deforestation	(+) 49 thousand ha of land is freed to other uses

* Sugarcane: 335 M litres/year x 1t/85 litres x 1 ha/70 tons = 56 thousand ha
Corn: 1 M l/d x 335 d/year x 1t/400 l x 6 tons/ha = 140 thousand ha

** In Brazil 2/3 of corn is planted as 2nd crop, sharing the land use with soybeans, therefore only half of the land is to be considered

*** Eucalyptus: 1 M litres/d x 335 d/year x 7.4 tons of wood/l x 500 tons of wood/ha = 5 thousand ha/yr x 5 years = 25 thousand ha of plant area of eucalyptus

**** 335 M l/year x 1 t/400 l x 0.28 t DDG/t = 235,000 t DDG/year
235,000 t DDG/year x 1UA/1.1 t DDG/year x 1 ha/head of animal = 214 thousand ha

4.3.4. SUGARCANE WITH CORN (FLEX DISTILLERIES)

As discussed in the previous sections, corn does have advantages but also disadvantages when compared with sugarcane as feedstock to produce fuel ethanol.

A comparative analysis of such the advantages and disadvantages are given in Table 4.5.

Table 4.5 Comparative advantages and disadvantages of corn and sugarcane as feedstock to produce ethanol. Source: composed by the authors.

Parameter	**Sugarcane**	**Corn**
Crop cycle	Semi-perennial crop Once planted, occupies the soil during 5 cuts, or ~7 years	Annual crop Can be planted as 1st crop or 2nd crop. Crop cycle ~4 months
Plantation/plant implementation time	~4 years	~1.5 years

Parameter	Sugarcane	Corn
Productivity	~7,000 litres of ethanol/ha/yr in Brazil	~4,000 litres/ha/yr in the USA ~2,500 litres/ha/yr in Brazil (crop 2nd crop)
Feedstock price volatility	Relatively low. Subject to be influenced by sugar demand	Relatively high since corn is in greater demand
Investment in agricultural machinery	High (needs to harvest large volumes per hectare)	Low (harvesting of small volumes per hectare)
Capacity to store raw material Raw material highly perishable. Needs to be processed within ~24h Can be stored and used all year around		
Capacity to be transported	Sugarcane is composed by 70% water, therefore severe difficulties to be transported limiting logistics	Low water content. Corn can be easily transported to long distances
Industrial conversion	Relatively low. ~80 litres of ethanol/ton of sugarcane stalks	High. ~425 litres of ethanol/ton of corn
Investment in industry to process/convert	Requires bigger equipment because season is shorter	Processing season can last all year long
	High (heavy equipment for crushing)	Medium (needs extra tanks for saccharification but uses the same equipment after that)
Extra fuel	Large quantity of fibre (bagasse) is produced. Ratio 2/3 for sugarcane, ¾ for energycane	Low fibre. Needs extra fuel. In USA natural gas. In Brazil eucalyptus if only corn is used as feedstock
Residues utilization	Very low protein in the cane. Extra bagasse is typically used for energy in the mill. Electricity can be exported (cogeneration)	DDGW or DDGS can be produced and used for beef cattle feeding

Maybe a good solution could involve integrating a corn ethanol plant with an existing sugarcane sugar mill, as explained bellow. Typically, all the after-harvest corn dry matter remains in the field and is incorporated as organic matter during the tilling process for soybeans planting. There is a controversy about how much corn stover can be left on the field and how much should be removed.

Flex distilleries is a new concept developed in Brazil using both feedstocks, sugarcane, and corn. This allows taking maximum advantage of both feedstocks, as indicated in Table 4.2 above.

Flex distilleries operates all year round, with sugarcane plant operating from April to October. The plant then saves bagasse for corn operation from November to March. Several flex mills report operation of 6 months for each feedstock.

Typically, a corn annexed distillery is built alongside an existing sugarcane ethanol plant. Normally this occurs in existing sugarcane plants built by traditional producers of sugarcane from Central South Brazil.

In addition, projects mentioned in the Table 4.1, three other flex projects are currently being constructed in Goiás.

Besides Mato Grosso and Goiás, there is also a good potential to install corn ethanol distilleries in São Paulo state where more than a hundred mills are already in operation and where there is more than 9 Mha of pastureland for beef cattle. São Paulo state offers an excellent opportunity for flex distilleries because of potential use of DDGW to promote beef cattle integration. Around 5% of beef cattle is produced in feedlots, and large part of this already occurs in São Paulo state.

A traditional sugarcane equipment manufacturer, the DEDINI company, is already installing flex distilleries in Brazil.[27]

FINAL REMARKS

As can be appreciated from previous pages, Brazil has considerable R&D capacity and know-how to develop new feedstocks, particularly from woody biomass, and process technology. Significant R&DD has been underway for many years in universities, research institutions and industry. Many routes are being investigated and it can be said it is an invigorating and promising research field.

With the right policies and support, this area could advance rapidly. This chapter has looked at potential and realistic models and strategies, taken into consideration environmental sustainability. It has considered process technology for new feedstocks and problems posed by land use and food production. It is a rapidly evolving field. It is also clear that there is considerable room for improvement, away from current unsustainable economic and environmental models.

For example, a better management of cattle ranches can easily result in freeing considerable amount of land while being environmentally more friendly. The combined production of sugarcane-corn presents an excellent social, economic, and environmental benefits while making better use of existing natural resources.

These are, therefore, realistic alternatives to current unsustainable production models, ranging from process technology to develop sustainable feedstocks, to economic models that make better use of existing resources with considerable economic, social, and environmental benefits.

27 https://www.dedini.com.br/index.php/noticias-eventos/noticias/92-usinas-flex-cana-milho-e-destilaria-modular-dedini-de-etanol-de-milho

REFERENCES

APLA. (2014). *Etanol 2G já está sendo produzido no Brasil.* Available at http://www.apla.org.br/etanol-2g-ja-esta-sendo-produzido-no-brasil.

BNDES. (2008). *Sugarcane-based bioethanol*: energy for sustainable development / coordination BNDES and CGEE, Rio de Janeiro, Brazil, 304 p.

CONAB – Companhia Nacional de Abastecimento. (2022). *Séries históricas.* Cana-de-açúcar, CONAB, Brasília. Available at https://www.conab.gov.br/info-agro/safras/serie-historica-das-safras?start=20

CORTEZ, L. A. B. (org.) (2016). *Proálcool 40 anos*: universidades e empresas: 40 anos de ciência e tecnologia para o etanol brasileiro. Blucher, São Paulo, Brazil, 224 p.

CORTEZ, L. A. B. Impact of corn ethanol production in land use in Brazil, In: *CRC Press Handbook of Bioethanol Fuels* (in Press).

EPE – Empresa de Planejamento Energético. (2022). *Plano Decenal de Expansão de Energia-2031.* Ministério de Minas e Energia, EPE, Rio de Janeiro, 411 p.

EQUIPAV. (2009). *Uso da palha da cana-de-açúcar.* EQUIPAV, São Paulo.

G1. (2015). *Novas usinas de etanol 2G no Brasil vão custar R$ 2,5 bilhões até 2024,* by ASSEMCIO C. Available at https://g1.globo.com/sp/piracicaba-regiao/noticia/2015/07/novas-usinas-de-etanol-2g-no-brasil-vao-custar-r-25-milhoes-ate-2024.html#:~:text=Novas%20usinas%20de%20etanol%202G,2024%20%7C%20Piracicaba%20e%20Regi%C3%A3o%20%7C%20G1&text=Al%C3%A9m%20da%20f%C3%A1brica%20inaugurada%20em,se%20igualar%20a%20combust%C3%ADvel%20comum.

IPAM – Instituto de Pesquisa Ambiental da Amazônia. (2016). *Desmatamento nos Assentamentos da Amazônia* – Histórico, Tendências e Oportunidades. Brasília, DF, Brazil, 114 p. Available at https://ipam.org.br/wp-content/uploads/2016/02/Desmatamento-nos-Assentamentos-da-Amaz%C3%B4nia.pdf.

KIT/IPT. (2022). *20º Webinar COGEN e única*: Cenários no mercado de energia elétrica. Presentation made by Caroline Schmitt (KIT) and Renata Moreira (IPT), 2022. Available at https://www.youtube.com/watch?v=xB_GvHqZw9U&t=2s.

LEITE, R. C. C. (coord.) (2009). *Bioetanol Combustível*: uma oportunidade para o Brasil. (in Portuguese), CGEE, Brasília, DF, Brazil, 536 p.

NASSAR, A. M.; RUDORFF, L. B.; ANTONIAZZI, L. B.; AGUIAR, D. A.; BACCHI, M. R. P.; ADAMI, M. (2008). Prospects of sugarcane expansion in Brazil: impacts on direct and indirect land use changes. In: ZUURBIER, P.; Van de VOOREN, J. (eds.). *Sugarcane Ethanol*, Wageningen Academic Publishers, Wageningen, The Netherlands, pp. 63-93.

OGEDA, T. L.; PETRI, D. F. S. (2010). *Hidrólise Enzimática de Biomassa.* Química Nova, v. 33, pp. 1549-1558.

PEREIRA, M. T. J. A. (2019). *Etanol de 2ª Geração*: expectativa e desafios. (In Portuguese), Undergraduate Project, School of Engineering, University of São Paulo-USP, Lorena, SP, 42 p.

RAÍZEN. (2022). *Sustainability of Raízen´s Ethanol*: ethanol production from sugarcane for thriving and sustainable economy. Raízen Technical Report prepared by Raízen with advisory services from Agroícone and verification by PwC (PricewaterhouseCoopers Independent Auditors), 66 p.

RFA. (2022). *Corn Ethanol's Energy Balance Continues to Improve*. Renewable Fuel Association-RFA, April 2022. Available at https://ethanolrfa.org/file/2214/Ethanol%20Energy%20Balance%20Update%20April%202022.pdf.

SANTOS, F. A.; DE QUEIRÓZ, J. H.; COLODETTE, J. L.; FERNANDES, S. A.; GUIMARÃES, V. M.; REZENDE, S. T. (2012). Potencial da palha de cana-de-açúcar para produção de etanol, *Quim. Nova*, Vol. 35, No. 5, 1004-1010.

SEARCHINGER, T. et al. (2008). Use of US Croplands for Biofuels increases Greenhouse Gases Through Emissions from Land-Use Change. *Science* 319(5867):1238-1240.

SOUZA, A. de N. de. (2010). *Estudo das Demandas de Etanol e Gasolina no Brasil no Período 2001 – 2009* (in Portuguese). Master Professional Dissertation. Escola de Economia de São Paulo, Fundação Getúlio Vargas, São Paulo, Brasil, 61 p. Available at https://bibliotecadigital.fgv.br/dspace/bitstream/handle/10438/8324/66080100274.pdf.

UNICA. (2022). *Bioeletricidade em números-ano base 2021*. 7 p. Available at https://observatoriodacana.com.br/listagem.php?idMn=134

CHAPTER 5
THE FUTURE OF BIOFUELS IN BRAZIL

5.1. INTRODUCTION: RECENT CHANGE IN HABITS AND POSSIBLE IMPACTS IN TRANSPORTATION IN BRAZIL

In recent years, with the advent of the Covid-19 pandemic, have witnessed significant changes in consumer habits particularly in middle and high-income countries. This trend goes back to recent decades, when a change in the perception of vehicles ownership has been accentuated, specially within the new generation. This change needs to be better understood as it has wider implications and embraces owning a car or a house, together with the diversification of transport modes of individuals. For example, the increasing use of motorcycles, bicycles, scooters, and variations of these modes, including electric, are visible even in middle income countries such as Brazil. Probably, this change in habits is partly due to concern with the environment and climate change.

Brazil could be considered as paradise for motorcycles and their manufacturers e.g., YAMAHA, HONDA, and others. Most of these motorcycles are used in delivery services in the middle and large cities in the country, making the traffic even worse, particularly in the rush hours and noon.

Also, as a complement, mobility companies such as UBER have gained importance during the pandemic. For example, in Brazil and particularly in large cities, it is frequent the use of their services for ride-hailing, food delivery, couriers, and freight transportation.

Besides, ecommerce is another kind of service that is gaining a significant momentum, particularly during the pandemic. Also, in Brazil, with a well-established domestic aviation network, local companies such as MARISA, MERCADO LIVRE, and AMERICANAS, are now disputing this market for delivery and ecommerce.

Finally, the pandemic accelerated another important change, homeworking, in detriment of office working. This has major implications e.g., for transport and shops that depend on the commuter for their business. Brazil has been one of the countries that suffered the most with the pandemic forcing many companies to adapt to changing consumer's preferences. Another important noticeable change has been the non-presential meeting platforms in favour of video conferences, which has also resulted a substantial change in habits. It is difficult to ascertain, but one can guess that this "new normal" will have an impact on transport, and consequently on liquid fuels, particularly in light vehicles sector in Brazil. Homeworking will not go away, although it is expected that the return to the office will gradually increase.

5.2. INTRODUCTION OF ELECTRIC CARS IN BRAZIL

Electric cars are gradually entering the Brazilian market, with multinational corporations plying a key role. According to the Brazilian Association of Electric Cars (ABVE), 801 electric cars were sold in 2020 and 2,851 in 2021. This figure represents near 1.5% of light vehicles sold in 2021 (1,557,957 and 416,474 commercial light vehicles) according to the Brazilian National Federation of Vehicles (FENABRAVE). These figures do not include other vehicles such as battery-electric vehicles (BEV), hybrid electric vehicles (HEV), nor plug-in hybrid electric vehicles (PHEV).

Table 5.1 Ranking of best-selling electric cars (EV) in 2021 in Brazil. Source: Canaltech (2021).

Model	Units sold	Price R$	Price US$[1]
Nissan Leaf	439	287,300	56,330
Porsche Taycan	379	699,000	137,580
Volvo XC40	375	389,950	76,460
Mini Cooper Electric	313	239,990	47,060
Audi E-Tron	252	604,900	118,600
BMW i3	159	304,900	59,780
Fiat 500e	146	248,000	48,630
Chevrolet Bolt	132	275,100	53,940
JAC iEV	49	169,900	33,310
Renault Zoe	53	204,000	40,000

As for prices to the consumer, it is difficult to compare the retail prices of electric with combustion engine vehicles, because they use different technologies with different maturity.

1 Using average exchange rate of R$ 5.10/US$ for 2021.

If the BEV, HEV, and PHEV, are added, the total number of electric cars sold in 2021 was 34,990 representing a rise of 77% over 2020.

Table 5.2 shows the ranking models of EV according to ABVE, 10 BEV, HEV, and PHEV sold in Brazil in 2021

Table 5.2 Ranking of BEV, HEV, and PHEV vehicles sold in in Brazil in 2021. Source: ABVE (2021).

Model	Units sold
Toyota Corolla XRX Hybrid-HEV	11,027
Toyota Corolla Altis-HEV	7,921
Volvo XC60-PHEV	3,366
Volvo XC40-PHEV	3,067
Volvo XC90-PHEV	982
BMW X3 XDRIVE 30-E-PHVE	836
BMW X5 XDRIVE 45-E-PHVE	812
Toyota RAV4H-PHEV	810
BMW 330E-PHVE	579
Porsche Cayenne PHVE	554

According to ABVE the number of BEV, HEV, and PHEV vehicles sold in the first trimester of 2022 alone, has increased by 115% compared with the same period of 2021. However, the perception of consumers in Brazil is that they will need to have significant reduction in costs to increase their share of the automobile market. This cost reduction in electric vehicles may occur as the sales accelerate in China, Europe, and the USA. This electric vehicle cost reduction process will determine the speed at which electric cars will be introduced and the impact on the Brazilian market too.

Finally, the future of fuel ethanol in Brazil will certainly depend on the price of electric cars in the domestic market. This transition will take at least two decades, and in the meantime, it will be necessary to establish specific policies to protect the biofuels market given their socio-economic and political implications for the country.

Another important difficulty associated with the introduction of electric vehicles in Brazil is lack of infrastructure to recharge these electric cars. Currently the country has more than 40,000 gas/petrol stations with regular ethanol (E100) and gasohol (E27.5) supply.

Although the introduction of electric cars maybe seems far into the future, electricity utility companies such as Companhia Paulista de Força e Luz (CPFL), today controlled by Chinese capital, are already installing free electric recharging stations. In 2015 the Chinese company BYD, manufacturer of electric buses, vans, trucks, and vehicles, inaugurated a new industrial plant in Brazil. Several BYD electric buses can already be seen circulating in Campinas. More recently, in 2022, the Chinese company Great Wall Motors (GWM) announced the installation of 100,000 electric vehicles factory, also in the same area. Therefore, it seems that there is already a strategy to introduce electric cars in the Brazilian market and consequently it is quite possible that this market will accelerate more rapidly than appears today.

A good example is the CPFL directly involvement in the electric car business. An important question is: where is the electricity coming from? In Brazil 85% of its electricity is generated from RE, primarily hydro power which is still the key. However more renewable electricity will be needed, and Brazilian electricity generation system needs to be properly adjusted and updated to accommodate the new demand. This will be challenging.

Naturally, more electricity will be needed, and that RE energy (mainly solar, wind, and biomass) are presently the best candidates for a rapid expansion. Wind has an important niche in Brazilian Northeast where already provides more than half of all electricity consumed. Biomass, in particular sugarcane mills, all together represents nearly 10% of all electricity consumed in the country. The share of solar is still very small but growing fast in a country with excellent sunshine potential. Solar will certainly benefit from cost reduction as it is happening around the world.

Already these three sources are providing almost 25% of Brazilian electricity. They offer the possibility of decentralized generation to enhance the security of the electricity grid. Therefore, there are good opportunities to significantly expand renewable energies, particularly for electricity generation.

5.3. PERSPECTIVES FOR FUEL ETHANOL

As a result of oil crisis in the 1970s Brazil intensified its production and use of sugarcane fuel ethanol, as discussed. During the past few decades, the country has built up a solid contribution of fuel ethanol to its liquid fuel energy matrix. However, this situation is under threat by the shift to electric light vehicle transportation technology. The introduction of electric vehicles, as indicated in Chapter 1, is growing a phenomenal speed, although unequally worldwide. China and Europe are cases in mind, but there remain serious unanswered questions as to where and how electricity will be generated. Will renewables be sufficient? Probably not and hence new alternatives need to be found.

However, as mentioned in Chapter 1, the future global energy scenarios presented by the International Energy Agency, BP, etc., forecast an important and growing contribution of modern bioenergy and advanced biofuels. Besides, difficulties related with the development of better batteries creates a situation in which liquid biofuels may still be a good solution for long distance transportation in a wide range alternative e.g., vehicles, aviation, and maritime, and the chemical sector. Table 5.3 shows different alternative fuels and their respective energy density.

Table 5.3 Different alternative fuels and their respective energy density: Sources:[2]

Transportation mode	alternative	energy density
Light vehicles	lithium-ion batteries	0.875 MJ/kg[3]
	anhydrous ethanol	30 MJ/kg
	gasohol E10	43.5 MJ/kg
Aviation	jet fuel - kerosene	35 MJ/l
	liquid hydrogen	10 MJ/l HHV 8.5 MJ/l LHV
	anhydrous ethanol	24 MJ/l
Maritime	diesel fuel	45.6 MJ/kg

Obviously, energy density is just one but important criteria in the selection of an appropriate fuel. And here biofuels may have an important niche to replace fossil fuels in long distance transportation. Also, an obvious question would be what is an ideal fuel? Could it be developed? Should it fit in the existing engines? Or should new engines be designed to fit an ideal fuel (s)?

5.3.1. THE FLEX-FUEL ENGINES IN BRAZIL

In the past two decades the most significant innovation of the automotive fuel use in Brazil was the introduction of flex-fuel engines with the ability to accept different proportions of ethanol and gasoline blends.

Flex-fuel vehicles were a positive answer to allow greater flexibility to consumers. The 1989 crisis was very disappointing for lack of planning, but the introduction of flex-fuel vehicles provided the necessary confidence on ethanol fuel. However, the flex-fuel engines were not a long-term satisfactory solution because the energy efficiency stagnated over time, creating the impression that light vehicles in Brazil were inefficient. This chronic problem, inherent to flex-fuel engines, together with a global effort to reduce GHG emissions, has created the momentum to revise policies and to consider changes to improve efficient mobility.

Consumers are not really satisfied with low performance of the flex-fuel engine. Table 5.4 illustrates fuel consumption of flex cars for E27.5 gasoline versus E100. The natural consequence is that the market for E100 is declining.

2 https://en.wikipedia.org/wiki/Energy_density

3 https://en.wikipedia.org/wiki/Lithium-ion_battery

Table 5.4 Fuel Consumption (gasoline E27.5 and E100) and cost per km for different flex-fuel vehicles in Brazil 2021-2022. Source: SODRÉ (2022).[4]

<table>
<tr><th rowspan="2">Vehicle Model</th><th colspan="4">Urban Fuel Consumption</th><th colspan="2">Fuel Price</th><th colspan="2">Cost per kilometer</th><th rowspan="2">(%) Difference E100 and E27.5</th></tr>
<tr><th colspan="2">Gasoline E27.5 (km/l)</th><th colspan="2">Ethanol E100 (km/l)</th><th>Gasoline E27.5 (US$/l)</th><th>Ethanol E100 (US$/l)</th><th>Gasoline E27.5 (US$/km)</th><th>Ethanol E100 (US$/km)</th></tr>
<tr><td>Hyundai Creta 1.6 flex 2021</td><td>12.3</td><td rowspan="2">0.8%</td><td>8</td><td rowspan="2">12.5%</td><td>1.12</td><td>0.86</td><td>0.091</td><td>0.108</td><td>+ 18.7</td></tr>
<tr><td>Hyundai Creta 1.0 turbo flex 2022</td><td>12.4</td><td>9</td><td>1.57</td><td>1.11</td><td>0.127</td><td>0.123</td><td>− 3.1</td></tr>
<tr><td>Honda City 1.5 flex 2021</td><td>11.1</td><td rowspan="2">4.0%</td><td>8.5</td><td rowspan="2">1.1%</td><td>1.12</td><td>0.86</td><td>0.101</td><td>0.101</td><td>0</td></tr>
<tr><td>Honda City 1.5 flex 2022</td><td>12.9</td><td>8.6</td><td>1.57</td><td>1.11</td><td>0.122</td><td>0.129</td><td>+ 5.7</td></tr>
<tr><td>Jeep Compass 2.0 flex 2021</td><td>8.1</td><td rowspan="2">33.3%</td><td>6.4</td><td rowspan="2">28.1%</td><td>1.12</td><td>0.86</td><td>0.138</td><td>0.134</td><td>− 2.9</td></tr>
<tr><td>Jeep Compass 2.0 flex 2021</td><td>10.8</td><td>8.2</td><td>1.57</td><td>1.11</td><td>0.145</td><td>0.135</td><td>− 6.9</td></tr>
</table>

Note: fuel prices for October 2021 and April 2022; price increase for gasoline E27.5 was 40.2% and for ethanol E100 was 29.1% – Exchange rates: October 2021 (1 US$ = R$ 5.65) and April 2022 (1US$ = R$ 4.60)

Light vehicles manufactured in 2022 are more efficient and less pollutant and more economic, reflecting, according to Sodré (2022), the improvements as a result of the Program for Vehicle Emission Control (PROCONVE)[5] coordinated by the Brazilian Ministry of Environment. All the participating vehicle models show an improvement in their fuel performance.

Although hydrous ethanol emits much less CO_2, consumers tend to be pragmatic and choose E27.5. The question is what will happen with E100 and flex-fuel engines in Brazil? Will simply stop producing and distribute E100? Can the existing infrastructure be used for hybrid or fuel cells technology using E100?

The 2021-2022 improvements in vehicle performance were not sufficient to offset the increase of fuel prices due to the Russian war on Ukraine despite better exchange rate in Brazil. As a result of the increase in oil prices, commodity prices (agriculture and iron ore) also went up. This phenomenon made the Brazilian Real recover much of its devaluation since the beginning of the pandemic in March 2020. Presently, the

4 SODRÉ, E. "Gasolina cara reduz vantagem de carros novos mais econômicos", in Folha de São Paulo, April 22th, 2022, Mercado, Page A12.

5 https://www.in.gov.br/materia/-/asset_publisher/Kujrw0TZC2Mb/content/id/56643907

price of gasoline (E27.5) is high compared with other nations, although it is experiencing fluctuations due to a new policy of tax reduction on liquid fuels.

Table 5.5 Price of gasoline in different countries. Source: BALAGO (2022).[6]

Country	Price (R$/l)
Argentina	4.76
Mexico	5,48
United States	5.59
Chile	6.50
China	6.91
South Africa	6.95
India	6.99
Brazil	7.19
France	8.85
UK	9.93
Germany	10.05

Why ethanol prices went up following the gasoline price increase? Maybe because April prices are traditionally high due to lack of ethanol in the market. Sugarcane harvest begins in April and May in Central-South Brazil. Another explanation maybe the opportunism of ethanol producers trying to extract extra earning.

Typically, ethanol (E100) prices in gas stations are about 70% of gasoline (E27.5) prices. Consumers use a rule of thumb according which if ethanol (E100) prices are higher than 70% of gasoline E27.5 price, they prefer to use gasoline (E27.5) instead of ethanol E100.

According to the above table values, in October 2021 ethanol E100 price was 76.8% of the price of gasoline E27.5. In April 2022 ethanol price went back to 70.7% of the gasoline price.

2021 was particularly a bad year for sugarcane due to prolonged drought and the need to honour sugar exports, causing lower ethanol supply.

5.3.2. WHAT WILL HAPPEN TO E100 AND E27.5?

As mentioned in Chapter 2, Brazil has created a unique blending market for this fuel. This practice originated in the early decades of the 20th century, gained maturity

6 BALAGO, R. "Petróleo caro deve ajudar economia do Brasil em 2022, diz FMI", in Folha de São Paulo, April 20th, 2022, Mercado, A21.

during the years between WW II and the 1st oil shock in 1973. When PROALCOOL was created in 1975, the goal was immediately implemented to increase the percentage of ethanol blended in gasoline. Annexed ethanol distilleries were built alongside existing sugar mills. Later, after the 2nd oil shock of 1979 a new target was to introduce the E100 market, producing ethanol on autonomous distilleries. Also, an agreement was signed between the government, the automakers, and ethanol producers with the objective to abolish possible royalties associated with the development of E100 engine. This agreement supported by ANFAVEA was considered a turning-point on the development of fuel ethanol.

The next decade, the 1980s, was still marked by subsidies to ethanol. As it was explained before, Petrobras (the State Oil company) had an "ethanol account" which paid a higher price for ethanol than sold in petrol stations. The Brazilian Government was financing this account contributing to the country growing inflation. The first few years of 1980s was also marked by an increase on international interest rates and the Brazilian foreign debt. These pressures contributed to the fall of the military regime and consequently the beginning of a democratic government in the country. Under democracy, the ethanol account was difficult to maintain, exerting greater pressure on ethanol producers. This situation, together with the increasing sales of ethanol cars, culminated in the 1989 crisis. That year the Brazilian Government had to import methanol to satisfy the ethanol deficit in the gas stations. This fact created a great unhappiness of consumers with ethanol vehicles (E100) nicknamed "duck cars". E100 cars sales dropped fast as consequence.

The 1990s started with this new reality in which subsidies to ethanol were removed and producers had to work out their own financial stability. These years were marked by the expansion of sugar production and in less than 10 years Brazil become the largest sugar exporter in the World. Sugar paid well and the model sugar-ethanol production was consolidated in the 1990s. However, the ethanol marked suffered a drawback as the E100 cars production practically went to zero in this decade.

The first decade of 21st century started with a well consolidated sugar-ethanol sector but with doubts of the Brazilian automobile sector. The advent of flex-fuel engine by Ford and their introduction in Brazil represented a good opportunity to regain confidence and enthusiasm on E100 fuel. Consumers, once sceptical of the E100 vehicles, could now regain new confidence with the flex-fuel cars. However, there was a price to be paid, and that was the decrease in engine efficiency. Consumers were happy because they paid less for the fuel and have the warranty that in case there was a shortage of ethanol, they could still run their cars on petrol. The flex-fuel cars were so successful that several manufacturers decided to stop producing gasoline vehicles and manufactured exclusively flex-fuel versions.

Another important point is that the introduction of flex-fuel cars reactivated the E100 market e.g., the E100 market (hydrous ethanol of around 93-94% by volume). So, now with the flex-fuel cars there was much more security for the consumers and the government with the risks associated with oil price fluctuations and commodities such as sugar, for instance. For further details see Cortez et al (2014b).

***

With current high prices of fuels, including ethanol E100, there is a consumer's mistrust with the overall vehicle the engine performance in the Brazilian market. Consumers are aware that in the developed nations, the automakers are selling cars with have better engine efficiency and this is a source of frustration in Brazil. As it will be discussed further below, the Brazilian Government is aware of the need to improve engine efficiency for light vehicles and the creation of ROTA2030 program has this objective in mind. In addition, there exist an unwillingness of the sugar-ethanol sector to expand its production capacity because of the uncertainties with the Brazilian economy. Ethanol production has grown very slowly, if not stagnated, in the last decade. The only exception is corn ethanol which is growing quickly and already represents around 10% of ethanol produced in Brazil. Therefore, the question with the ethanol market would be what to do with the E100 fuel and flex-fuel cars? Would it be a good idea simply stop producing flex-fuel cars and gradually turn hydrous into anhydrous ethanol to supply more ethanol for blending? The recent USA increase of ethanol blending to B1[7]5 maybe is an indication of how ethanol can be part of the strategy to mitigate GHG emissions.

5.3.3. TECHNOLOGICAL ALTERNATIVES FOR FUEL ETHANOL

Although 1G biofuels, in many aspects, still have their comparative advantage in Brazil, the transition is posing new challenges as well as opportunities faster than we can imagine. The first possibility is to introduce technologies that can gradually introduce electrification combined with ethanol. This could take advantage of the existing E100 infrastructure and existing flexibility in light vehicle fuel market. According to Federal Government's Annual Statistics 2021[8] the present number of gas stations has grown, with over 41,000and 180+ distributors, with BR, IPIRANGA, and RAÍZEN being the most important, employing more than half million workers. Practically in all gas stations E100 and E27.5 is available.

In 2021, TOYOTA introduced the flex-fuel Corolla Altis Hybrid[9] petrol-electric model. This can be viewed as a trial to provide greater fuel flexibility to the consumer and a hope that the hybrid technology can compensate the lower performance of flex-fuel engines. These vehicles tend to be more expensive, of course, because they incorporate both the combustion and the electric engines.

7 https://www.tomorrowsworldtoday.com/2022/04/25/biden-announces-e15-rule-change-poet/

8 https://www.gov.br/anp/pt-br/centrais-de-conteudo/publicacoes/anuario-estatistico/anuario-estatistico-2021

9 https://www.toyota.com.br/modelos/corolla/

Using a different strategy, focused on fuel cells, NISSAN is developing a new ethanol-hydrogen engine. The new model NISSAN Solid Oxide Fuel Cell - SOFC[10] will not need an electric battery because ethanol E100 is used to produce the required hydrogen. The technology allows E100 to be converted into hydrogen using ethanol reformers and then use the hydrogen in a conventional lithium-ion fuel cell. A prototype has already been tested and is scheduled to be in the Brazilian market quite soon.

The key feature of fuel cell technology is to escape from Carnot´s limiting efficiency. All combustion engines or cycles, including Otto, Diesel, Brayton, and Rankine, are subject to Carnot´s efficiency as upper limit. The fuel cell technology breaks with that paradigm since it is not a thermal but an electrochemical reaction involving no combustion. The expected overall efficiency is the product of the reformer and the fuel cell efficiency.

There is ample room for more efficient ethanol engines. For example, FIAT[11] was developing a high-efficient ethanol engine though it gave up in 2021. It is difficult to speculate the company decision, but the most likely reason was the way the company evaluated the perception of Brazilian consumers on the future of E100. Nonetheless, this could be a viable technical alternative which should be considered further. However, the Brazilian automotive industry, basically multinational corporations, are less willing to find a solution that will only be valid in Brazil. As mentioned before, this situation happened in 1979 with the ethanol-based engine. At that time the government intervened to sort out this resistance from the car manufacturers. But those were different times, when the Brazilian Federal Government was a military dictatorship and had more capacity to literally dictate policies.

5.3.4. HOW ETHANOL ENGINES COMPARE WITH THE ELECTRIC CARS?

Sugarcane ethanol produced in Brazil has a very low GHG emissions and can adequately compete with electric cars, especially if the electricity is not sustainable. Table 5.6 published by CNPEM (2018), compares different engine technologies using ethanol with flex-fuel using gasoline and ethanol.

10 https://quatrorodas.abril.com.br/noticias/nissan-e-ipen-renovam-parceria-para-pesquisas-de-veiculos-eletricos/

11 https://quatrorodas.abril.com.br/noticias/fiat-desiste-de-lancar-motor-1-3-turbo-a-alcool-super-eficiente-no-brasil/

Table 5.6 Comparison of different engine technologies using ethanol with flex-fuel using gasoline and ethanol. Source: CNPEM (2018).

Type of vehicle	Source of energy	Emissions[12] (gCO_2e/MJ)	Energetic efficiency[13] (MJ/km)	Energy Emissions gCO_{2e}/km	Total Emissions (energy+vehicle)[14] gCO_{2e}/km
Flex	Gasoline A	87.4	2.19	191	226
Flex	Gasoline C (E27)	73.8	2.19	162	196
Flex	Ethanol (E100)	20.8	2.19	46	80
Hybrid	Ethanol (E100)	20.8	1.64	34	74
Electric	electricity	55.5	0.72	40	96

Ethanol may not have a comparatively good energetic consumption as electric cars or fuel cell cars but has very good CO_2 emissions/km when compared with gasoline, and other available fuels, in general. Although fuel cells using ethanol were not considered in this study, they could also be a good alternative both for mitigating GHG emissions/km and to increase energy efficiency (MJ/km).

Also, according to an AEA[15] publication (AEA, 2017), Figure 5.1, ethanol can compete in equal terms with electric cars as far as GHG emissions/km is concerned.

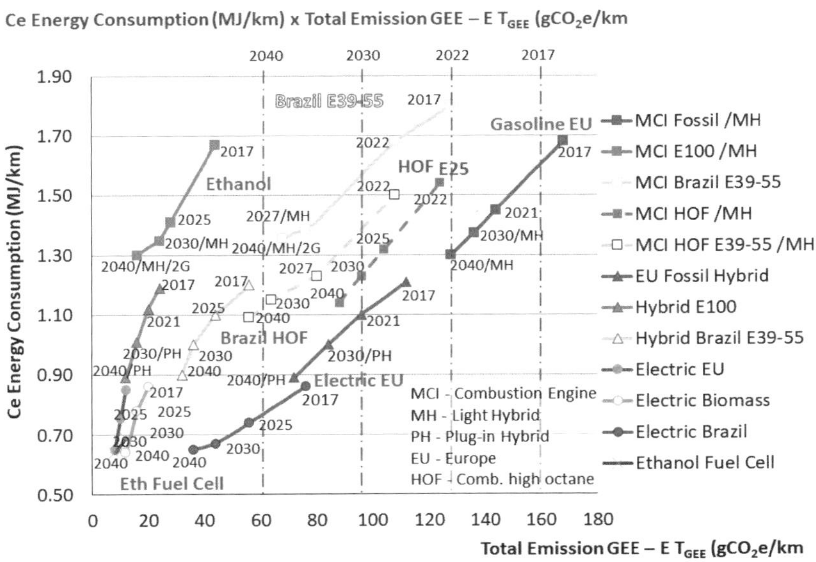

Figure 5.1 Energetic consumption vs total GHG emissions from well to wheel. Source: AEA (2017).

12 Based on RENOVABIO data

13 Based on Ecoinvent v3.1

14 Includes vehicle assembling and maintenance considering 288,394 km for a combustion engine vehicle, 216,296 for electric vehicles, and 100,000 km for the batteries

15 Associação Brasileira de Engenharia Automotiva – AEA (Brazilian Automotive Engineering Association)

Also, in the same direction, Figure 5.2 shows another interesting study, done by RAÍZEN (2022), comparing GHG emissions for electric passenger cars used in different countries and hybrid+ethanol in Brazil, plus ethanol in Brazil.

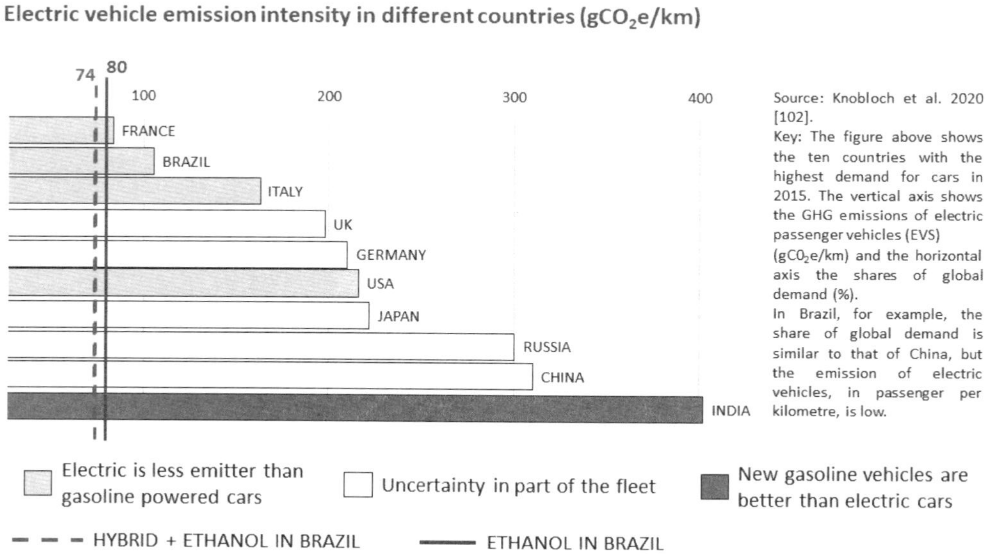

[12] The assumptions, parameters, and approaches adopted in the LCA can generate different results and interpretations for the individual transport technology.
The comparison between technologies can generate extremely heterogeneous and diverse results. This is mainly due to the different choices regarding the system boundary (border), level of detail in data collection (primary sources and aggregated data), assumptions and modelling [89]. Factors such as: vehicle performance, manufacturing, and type (materials, production process, weight, steel demand); technical characteristics (travel distance, charging time and schedule, efficiency, battery capacity), scope, time window, raw material extraction and production, final disposal, and other parameters can directly influence the results related to GHG emissions [258], [79].

Figure 5.2 GHG emissions (gCO2e/km) for electric passenger cars used in different countries compared with hybrid+ethanol in Brazil, and ethanol in Brazil. Source: (RAÍZEN, 2022).

And Table 5.7 shows also a RAÍZEN study comparing different electric car models with flex fuel engines using different fuels demonstrating their competitivity for GHG emissions.

Table 5.7 Comparative analysis of different electric car models (hybrid, hybrid plug-in, and electric) with flex fuel engines using gasoline A, gasoline C, and E100. Source: (RAÍZEN, 2022).

Region	Vehicle type	Vehicle's power source	Total (power and vehicle)
Brazil	Flex [83], [102]	Gasoline A	226
	Flex [83], [102]	Gasoline C [E27]	196
	Flex [83], [102]	Ethanol [E100]	80
	Hybrid [83], [102]	Ethanol [E100]	74
	Hybrid Plug-in [85]	Gasoline w/Electric	242
	Electric [83],[85]	Electricity	96-151

Region	Vehicle type	Vehicle's power source	Total (power and vehicle)
European Union	Electric [101]	Electricity	130
	Hybrid [89]	Gasoline	150-250
	Hybrid Plug-in [101]	Gasoline/Electric	90-190
United States China India	Electric [98]	Electricity	293
	Hybrid [106]	Gasoline	370
	Electric [107], [108], [86]	Electricity	91-265
	Gasoline [107]	Gasoline	199-250
	Electric [109], [103]	Electricity	289-400
Potential Technology	Hydrogen Gas [100]	Solar	92
	Hydrogen Gas [100]	Natural gas	450
	Liquid Hydrogen [100]	Solar	277
	Liquid Hydrogen [100]	Natural gas	660

Based on these data, it can be stated that there is no reason why sustainable ethanol production should not be expanded and used as an important component of the Brazilian GHG mitigation plan. It can be concluded that:

- Sugarcane fuel ethanol used in flex-fuel combustion engines already competes adequately in terms of the GHG emissions/km,
- The three other technologies (improve/dedicated E100, hybrid and fuel cells using fuel ethanol) could improve considerably the possibilities for ethanol.

Therefore, it could be concluded that the implementation and imports of electric models to Brazil is not only unnecessary but could be harmful.

5.3.4.1. Brazilian government policies for fuel ethanol

To meet with these needs, and comply with the Paris Agreement 2015, the Brazilian federal government approved in 2017 the RENOVABIO program to promote sustainable biofuels. In accordance with the Paris Agreement, Brazil accepted to reduce the GHG emissions by 37% in 2025 and 43% in 2030, relatively to 2005 levels.

Implemented by the Brazilian Agency ANP, RENOVABIO is the most recent National Biofuels Policy (*Law n. 13.576/2017*), with the specific objective to increase the production of sustainable biofuels in Brazil. RENOVABIO establishes targets to encourage biofuels market and reduce GHG emissions. It can be viewed as a transitional policy valuing ethanol and promote energy efficiency in transportation. RENOVABIO authorizes private companies to issue decarbonization credits known as CBios,[16]

16 CBio corresponds to one ton of carbon dioxide equivalent (tCO_2e)

to encourage renewable biofuels such as bioethanol, biodiesel and biomethane. The CBios are traded at the Sao Paulo stock market and in December 2021 its price was nearly US$10/t$CO_2$e.[17] Another complementary federal program is ROTA2030 (Law number 13.755/2018) which is intended for the automotive chain with the objective of supporting technological development, competitiveness, innovation, vehicle safety, environmental protection, energy efficiency and overall quality of automobiles. While RENOVABIO focuses on GHG emissions reduction, ROTA2030[18] focus is engine efficiency for example, reduction on GHG per km:

$$\underset{\left(\frac{CO_2}{litre}\right)}{RENOVABIO} \quad x \quad \underset{\left(\frac{litre}{km}\right)}{ROTA2030} \quad \left(\frac{CO_2}{km}\right)$$

Therefore, it can be concluded that the Brazilian federal government still remains an important protagonist in the automotive sector. The government is trying, somehow, "to run after the damage", as it is saying in Brazil, since there is a general feeling that ethanol as a fuel was about to be taken over by the electric car. Thus, it would be incorrect to say that, eventually, ethanol fuel will be replaced by electric vehicles, for both passenger cars and small commercial vans.

5.3.4.2. Complementary strategies ethanol with electric mobility

As indicated previously, imports of electric models (vehicles, motorcycles) would not be effective in terms of reducing GHG emissions for light vehicles. Besides, this policy could be risky because Brazil, being an atypical country in terms of natural resources, e.g., land, water, and sunshine, has excellent possibilities and opportunities to integrate biofuels and food, contrary to other countries where there could exist a conflict between food and biofuels. The math of GHG emissions is not always $1 + 1 = 2$. Sometimes $1 + 1 < 2$ as explained in chapter 4, with the introduction and expansion of corn ethanol. Having a prosperous agribusiness opens a tremendous opportunity for Brazil, since food is and will be the key most important item for human beings. And integrating the residual part of food production with the production of fuels may be a clever way to do so. This could be accomplished while achieving a substantial reduction of GHG in land change and beef cattle related emissions.

In addition, Brazil could, as it was discussed before, gradually adopt the electric cars in this transition without necessarily abandoning, sustainable fuel ethanol, from sugarcane or corn. This new strategy could be an effective way to take advantage of the climate change crisis, to promote sustainable economic development once again.

17 www.renovabio.org

18 https://www.rota2030.com.br/#:~:text=O%20que%20%C3%A9%20o%20Rota,e%20a%20qualidade%20dos%20autom%C3%B3veis.

5.4. PERSPECTIVES FOR BIODIESEL

Vegetable oils have been considered as diesel substitute since the 1970s. In a book published by Melo and Fonseca (1981) they state that the area required to substitute diesel would be much larger than to substitute gasoline with sugarcane ethanol. Actually, Brazil does not growth an energy crop to produce vegetable oil as good as sugarcane ethanol. While overall productivity for sugarcane ethanol is around 7,000 litres/ha/yr, for soybean biodiesel it is around 400 litres/ha/yr. With palm oil the result could be much better, but Brazil was never able to obtain good results with palm oil. There have been a few projects e.g., PETROBRAS and VALE, but so far failed to demonstrate any significant increase in agricultural yields.

In 2010, PETROBRAS informed of the construction of a biodiesel plant using palm oil in Pará, with investment both in agriculture and industry. With three others previously constructed plants (Montes Claros, MG, Candeias, BA, and Quixadá, CE), PETROBRAS planned to produce 108.6 million litres of biodiesel with soybeans as the main feedstock. Unfortunately, the original plans were not successful. PETROBRAS biodiesel from palm oil was a failure, due primarily to agronomic difficulties.[19] On the other hand, Petrobras green diesel and hydrotreated vegetable oil (HVO) is doing well at its President Getúlio Vargas Refinery (REPAR), in Araucária, Paraná. They process 2 million litres of soybeans oil and 40 million litres of diesel with renewable content.[20]

VALE also implemented BIOPALMA, a palm oil plantation and a biodiesel plant in Pará in 2012. This plant produces 25 tons of oil/h, and the main idea is to produce B20 (20% biodiesel + 80% mineral diesel). The B20 produced by VALE will have significant applications, including fuel for the iron ore trains from Carajás, Pará, to the Ponta da Madeira harbour in Maranhão State, about 900 km. Other possibility is to fuel large trucks in operation at the Carajás iron ore mine.

Again, the main challenge remaining is the sustainable production of feedstocks for biodiesel. It seems then inevitable the integration with food production, considering the economic advantages. Today nearly 78% of biodiesel produced in Brazil uses soybeans as feedstock. This is because soybean cake covers production costs, and therefore, biodiesel remains a marginal beneficial activity.

Although this food-biofuel integration is positive, it has its limitations since no significant expansion is foreseen, unless more soybeans cake can be produced and consumed. In 2021 started with B12, then increased to B13 for two months, and because of the Russian war with Ukraine, and corresponding commodity price increase, the Brazilian Federal Government reduced the blend to B10. This shows how difficult it is to maintain a certain blend based on commodity prices, although this seems to be a reasonable price to pay, and that flexibility and adjustment are necessary.

19 https://www.poder360.com.br/luanda-leaks/petrobras-enterra-centenas-dende/

20 https://petrobras.com.br/fatos-e-dados/concluimos-testes-para-producao-de-diesel-renovavel.htm

It can be stated that in the short-term the future of biodiesel in the country depends on soybean integration, therefore with a slow yearly increment. If greater and faster production is achieved, more efforts should be made to improve agricultural yields such as palm oil.

5.5. EMERGING MARKETS FOR BIOFUELS

5.5.1. AVIATION

Worldwide, in the transportation sector, aviation is the 4th most important emitter of CO_2 (Table 5.8 below):

Table 5.8 Distribution of CO_2 emissions in the transportation sector worldwide, in 2020. Source: Statista.[21]

Sector	%
Passenger cars	41
Medium and heavy trucks	22
Shipping	11
Aviation	8
Buses and minibuses	7
Light commercial vehicles	5
Two/three wheelers	3
Rail	3

As stated in Chapter 1, the interest in aviation Sustainable Aviation Fuels (SAF) has increased dramatically in recent years. Currently all major airlines are scrambling to find SAF, driven primarily by environmental considerations, but the main constrain remains the cost of sustainable feedstocks. Both IATA and ICAO have introduced specific policies to reduce emissions. In a few years, the vision and attitudes of air travel have shifted significantly in favour of achieving carbon neutrality, and despite high costs, SAF is increasing dramatically e.g., in 2021 production and use was estimated between 100 to 150 million litres, a 50% increase over the previous year. By 2030, projections are that SAF could provide at least 5% of global demand (IATA, 2021).

This projection could easily be overcome if present trends continue as all major airlines are investing in SAF e.g., investigating and testing new fuels. Advances in process technology are cutting costs considerably and opening the utilization of new feedstocks much faster than envisaged a few years ago. Although there is a long way to go since aviation requires a very high-quality fuel, the future looks much brighter!

21 https://www.statista.com/statistics/1185535/transport-carbon-dioxide-emissions-breakdown/

For example, British Airways has become the first airline in the world to use SAF on a commercial scale, after the company Phillips 66 Ltd delivered the first batch. The company is currently producing about 500,000 litres/day from waste feedstocks (Biofuels Inter. March 30/22)

Brazil is the world's 4th largest domestic air market. This is due to the country size combined to a lack of good road and railroad infrastructure. According to IATA (2019) with a gross value added of R$ 18.8 billion to the national economy and generating more than 800,000 direct and indirect jobs, the Brazilian aviation market has the potential to grow almost 5 times in 20 years. Airlines, airport operators, aircraft manufacturers and service providers employ about 167,000 people directly and over 372,000 indirectly.

The Brazilian aviation sector does not allow foreign companies to operate domestically although there is no restriction for international flights. The main private companies operating are currently LATAM, GOL, and AZUL, all involved in a common effort to use sustainable biofuels. Typically, they operate using BOEING, AIRBUS aircrafts, but EMBRAER, a Brazilian airplane manufacturer, is also an important player, particularly in the mid-size craft production.

For instance, GOL used biofuel blend with kerosene in their regular flight from Recife to Fernando de Noronha, a paradise island near the Brazilian Northeast coast.[22] During the 2010 decade, AZUL also made few experimental flights using sugarcane derived biofuel produced by AMYRIS.[23] The company SOLAZYME also produced sugarcane derived biofuel for aviation.

Brazilian airlines have demonstrated strong interest to promote the development and use of SAF. Naturally, the main idea is to comply with established targets to reduce GHG emissions but also for marketing and economic benefits.

Specifically for the aviation sector, the target is to have a carbon neutral after 2020, and a 50% reduction in GHG emissions up to 2050, compared with 2005 levels (FREIRE, 2011). The 41st Assembly of ICAO agreed to aim for zero CO_2 emissions by 2050 (see Chapter 1). Progress ais being made by the aviation sector to improve operations with higher load factors and utilizing bigger planes. Of course, the pandemic created difficulties, with less passenger flights, but the industry is still pursuing its goals.

The difficulties associated with biofuels for aviation are mainly economic. All major airlines such as Air France[24] recognizes that "SAF is currently much more expensive than fossil fuel". Most airlines are investigating and testing new fuels, but they must also comply with the American Society of Testing of Materials (ASTM). ASTM is the

22 https://www.gazetadopovo.com.br/economia/brasil-e-apenas-um-espectador-no-avanco-dos-biocombustiveis-para-avioes-1u4ee5hd6dx6x5s6a4brzc1oe/

23 https://www.biodieselbr.com/noticias/biocombustivel/bioqav/azul-gol-voos-experimentais-cana-acucar-oleo-milho-190612

24 https://wwws.airfrance.com.br/information/developpement-durable/sustainable-aviation-fuel

only authority to certify biofuels for aviation accepted by the insurance companies. The insurance companies follow exactly the recommendations established by the turbine manufacturers, so it prevails the "drop-in" approach in the development of biofuels for aviation.

Considering this premise, a roadmap for sustainable biofuels for aviation in Brazil was conducted in 2014 by BOEING/EMBRAER with support from FAPESP (BOEING/EMBRAER/FAPESP, 2013; CORTEZ, 2014a). The project analysed different possible pathways for biofuels for aviation with Brazilian feedstocks, and the complexity of established conversion industry. The drop-in approach was the main boundary condition in this project. The project made the following recommendations:

- Filling R&D gaps in the production of sustainable feedstocks,
- More incentives to overcome conversion technologies barriers including scaling up issues,
- Greater involvement and interaction between private and government stakeholders,
- Creation of a national strategy to make Brazil a leader in the development of aviation biofuels.

One conclusion of this study "is that the substation of petroleum in aviation represent a very important niche for sustainable biofuels. Brazil has a great opportunity in this area to become a global player. There are important challenges to be overcome to create the basis for this new emerging industry. Brazil cannot afford no participate".

Although there are today several pathways that have been certified by ASME, none satisfy both requirements: have a GHG emissions mitigation potential and reasonable costs to compete with KAV without any subsidy. Therefore, since the aviation sector needs to find a solution in the present decade (up to 2030) there're only three possible options today: subsidize the most reasonable pathways; compensation; or the non-drop-in approach.

Figure 5.3 summarizes the difficulties to mimic aviation kerosene, a specific fuel utilized in jet planes and used as reference for ASTM certification. According to the figure below, the best feedstock to produce a sustainable aviation biofuel remains vegetable oils such as soybean, camelina and palm oil. Note, they are close to the centre in the figure. As we come to the edge, the feedstock costs tend to decrease but the technical difficulties increase.

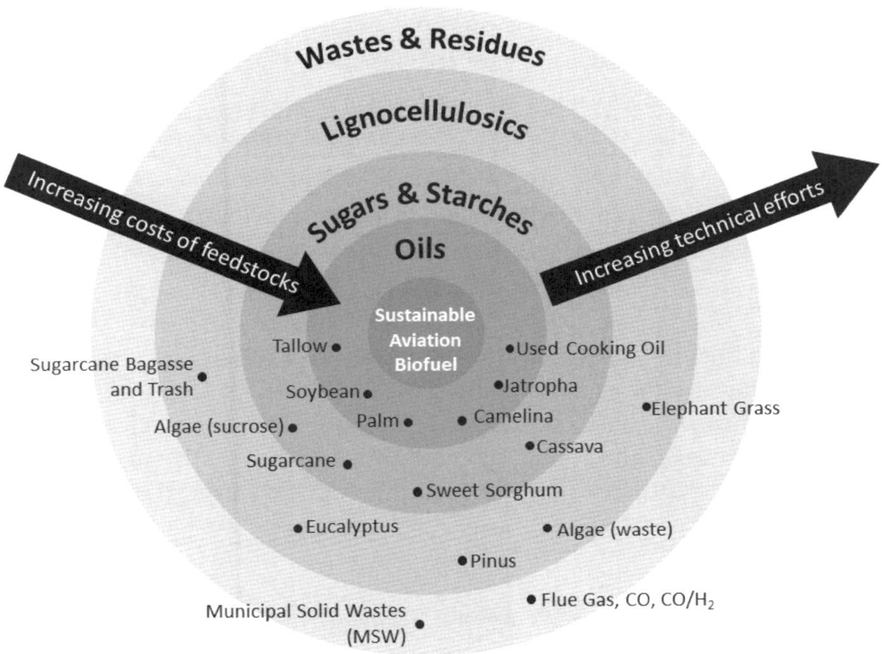

Figure 5.3 Raw materials and their relative position according to costs and technical efforts to be converted into sustainable aviation biofuel. Source: Cortez (2014a).

Brazil is one of few countries to have a specific policy on the use of biofuels in aviation. However, the initial strong motivation shown a decade ago, has decreased probably due to the economic crisis after 2012 and the difficulties associated with high production costs of biofuels for aviation.

In the BOEING/EMBRAER/FAPESP report several pathways for biofuels were analysed and are presented in Table 5.9 and Figure 5.4 below. As mentioned in the report, they present an overview of all identified pathways pertinent to Brazil, including the status of ASTM approval process.

Table 5.9 Identified and analysed pathways to produce biofuels for aviation in Brazil. Source: BOEING/EMBRAER/FAPESP (2013).

1	Oil-bearing plants through HEFA
1A	Tallow and fats through HEFA
1B	Used cooking oil through HEFA
1C	Sugar-bearing plants through fermentation of lipids and HEFA
1D	Ligno-cellulose through hydrolysis to sugars to lipids and HEFA
2	Sugar-bearing plants through fermentation to ethanol and ATJ
2A	Ligno-cellulose through hydrolysis to fermentable sugars to alcohol and ATJ
2B	Flue gas (CO, CO/H_2) through fermentation to alcohols and ATJ

2C	Ligno-cellulose through gasification to CO/H_2 to alcohols to ATJ
2D	MSW through fermentation to organic acids to alcohols to ATJ
3	Ligno-cellulose through gasification & reforming to syngas to FT
3A	Ligno-cellulose through pyrolysis to biochar &bio-oil to gasification to syngas to FT
4	Sugar-bearing plants though fermentation by GMO to hydrocarbons (DSHC)
4A	Ligno-cellulose though hydrolysis to sugars to hydrocarbons (DSHC)
5	Ligno-cellulose through fast pyrolysis to bio-oil via deoxygenation (HDCJ)
5A	Ligno-cellulose through direct liquefaction to bio-oil via deoxygenation (HDCJ)

According to the information above, up to June 2014 ASTM had already approved three biofuel pathways for jet fuel production: D7566 HEFA A2 which can use used cooking oil, oil-bearing plants, or tallow; D7566 SIP (DSHC) which can use MSW, starches, sugar-bearing plants, or flue gas; and D7566 FTA1 which can use algae or ligno-cellulose. Therefore, there is already the technology to convert the main feedstocks into aviation biofuel.

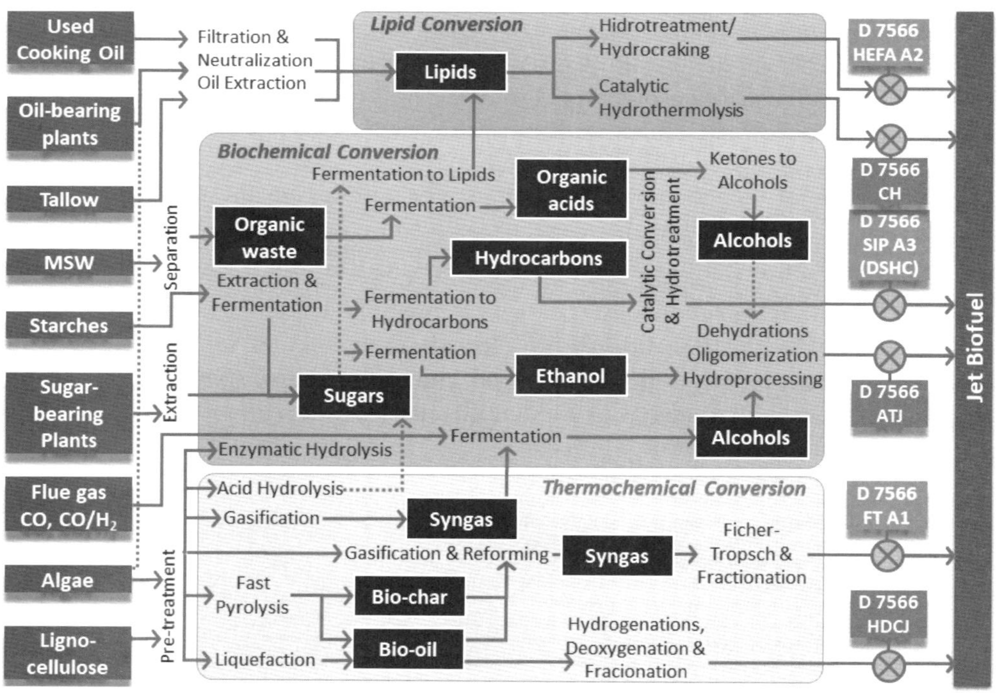

Figure 5.4 Identified pathways to produce sustainable jet biofuel in Brazil. Source: BOEING/EMBRAER/FAPESP (2013).

However, costs remain a critical factor. Costs to produce biofuels for aviation depend basically on two factors if a drop-in solution is to be pursued: i) the costs of the sustainable feedstocks and ii) processing technology (CORTEZ, 2014a). As seen above, in the drop-in approach, vegetable oils are typically closer to aviation kerosene (KAV) but are also expensive and compete with food and agriculture land. On the other hand, if a low-cost raw material is selected, the processing technology to comply with ASTM criteria will economically gain importance. From the GHG emissions mitigation potential, biofuel will heavily depend on the feedstock. Therefore, its correct identification is a crucial part of the process.

Although there are today more ASTM approved and certified pathways, none satisfy both requirements: have a GHG emissions mitigation potential and reasonable costs to compete with KAV without any subsidy. Therefore, since the aviation sector needs to find a solution in the present decade (until 2030) there're only three options today, as it was explained before.

Subsidy is a limited possibility and politically difficult solution. If necessary, airlines will accept subsidies only in few less expensive flights. The criteria may involve flights with commercial return such as tourist destinations. Broader subsidies are less likely to occur because it requires government participation. Considering the international nature of airlines, a possible alternative could be an international policy agreement.

Compensation is a good possibility from the GHG emissions mitigation point of view but is not a good option from the marketing point of view. There are several good technical opportunities for compensation. Brazil, for example, has around 200 Mha of pastureland, maybe 25% of that is either degraded or at high risk of degradation. Although, typically enterprises prefer a more direct way to reduce GHG emissions, LATAM has recently announced a new marketing strategy labelled "Sustainability: a necessary destination" (Folha, April 22nd, 2022) to compensate carbon emissions of its main routes on Fridays by means of projects to capture CO_2 emissions.

Finally, the non-drop-in approach could be an interesting way to address this problem. Instead of trying to look for an ideal and low-cost biofuel, the turbine companies could try to develop a turbine based on the characteristics of a low cost and sustainable biofuel.

5.5.2. MARITIME

Brazil has basically three markets for water transportation: long distance international maritime; coastal maritime; and fluvial transportation.

The long-distance maritime transportation is dominated by two companies, PETROBRAS for crude oil and its products,[25] and VALE, basically, for iron ore transportation. In addition to these two companies there are also large ships transporting agricultural commodities. Other maritime transport modalities such as containers,

25 Brazil exports crude oil and import its products. This happens because there is a lack of refining facilities.

passengers, and fishing are relatively small by comparison with the large bulk transport and their fuel requirements.

The ships employed among these companies are usually very large, often between 100-300,000 tons[26] transporting low-cost bulk materials such as oil, liquid fuels, iron ore, and grains. Therefore, the cost of maritime fuel cannot be excessive otherwise it makes the business uncompetitive. Although costal and fluvial transportation are lower than maritime, the same criteria (low-cost fuel) should apply.

Typically, the most utilized fuels in maritime transport sector are Heavy fuel oil (HFO), light fuel oil (LFO), and marine gasoil (MGO). The HFO, also named bunker oil, is the most common and consumed in high quantities, 250 million tons per year (RODRIGUE, 2020). All these fuels contain high amounts of sulphur[27] and are subjected to serious restrictions by The International Maritime Organization (IMO), a regulatory branch of the United Nations for the navigation sector. The IMO is committed to reduce the carbon impact of shipping by at least 50% compared to 2008 levels by 2050.[28]

Ongoing studies in Brazil are investigating the possibilities for fossil maritime fuels substitution, including sustainable biofuels. SZKLO et al. (2020) studied several fuels such as Straight Vegetal Oil (SVO), Hydrotreated Vegetable Oil (HVO), biomass FT-diesel, biomethanol (CH_3OH), and FT synthetic renewable H_2 and captured CO_2 (electrodiesel/e-diesel). In relation to GHG emissions, all studied fuels shown high mitigation potential (average 75%), with e-diesel close to 97%. As for production costs, HVO is the alternative with lowest cost followed by FT-diesel, biomethanol, and e-diesel. However, e-diesel production costs are almost five times higher than HFO.

HFO seems to be recognized as a promising biofuel for maritime applications. Since 2020, Petrobras[29] has been producing Hydrotreated Vegetable Oil (HVO) for certain applications.

Galindo et al. (2020) studied blends using sugarcane and eucalyptus pyrolysis with ethanol and diesel. The study emphasized the difficulties of high levels of water in the biomass bio-oil, although few maritime engines accept some water content. Typically, studied blends were in the order of 10% of bio-oil, 60% of ethanol, and 30% of diesel achieving properties close to maritime diesel with low flash point. However, blends of 65% of bio-oil, 10% of ethanol, and 25% of diesel showed better flash point, although with lower stability causing phase separation.

Wei (2021) also studied the following alternatives: liquefied natural gas, methanol, ammonia, biodiesel, SVO, HVO and HPO.

---

26 A valemax ship can transport up to 400,000 tons. http://www.vale.com/brasil/PT/initiatives/innovation/valemax/Paginas/default.aspx

27 The maritime transportation sector responds for about 15% of global sulfur emissions.

28 https://www.imo.org/en/MediaCentre/PressBriefings/Pages/06GHGinitialstrategy.aspx

29 https://petrobras.com.br/fatos-e-dados/petrobras-defende-concorrencia-no-segmento-de-biocombustiveis.htm

Table 5.10 shows a summarized comparison between possible feedstocks and their correspondent biofuel, including potential advantages and disadvantages.

Table 5.10 Potential Biofuels for Maritime Applications. SOURCE: Compiled by the authors.

Feedstock	Conversion Technology	Biofuel type	Possible Advantages	Possible Disadvantages
Sugarcane/Energycane (whole), Eucalyptus, Crop Residues (palm)	Pyrolysis	Bio-oil: Blended "as is" Blended after upgrade (catalytic reform) Blended after refinery (biocrude)	AGR, CEff, LC, GHG	Maybe need to upgrade
Sugarcane bagasse from existing mills			CEff, LC, GHG, no additional AGR	Dependence from sugar mills
Urban Wastes			CEff, LC, GHG, no additional AGR	Variable feedstock composition
All fibre above	Fisher-Tropsch	FT-diesel	AGR, CEff, LC, GHG, drop-in	
All feedstocks above	Gasification, biodigestion and reforming	Hydrogen	GHG	High cost, complex logistics
All feedstocks above	Haber/bio-hydrogen	Ammonia	GHG	High cost, complex logistics
Sugarcane/Energycane	Fermentation	Ethanol	AGR, CEff, LC, GHG, commercial technology	Needs more land
	Fermentation (ABE Process)	Butanol	AGR, CEff, LC, GHG, commercial technology	Needs more land
Eucalyptus	Hydrogenation of carbon dioxide	Methanol	AGR, CEff, LC, GHG	Needs more land
Palm Oil	Hydrogen catalysis	Hydrogenated Vegetable Oil-HVO	AGR, CEff, LC, GHG	Possible competition with Food
	none	Straight Vegetable Oil-SVO	AGR, CEff, LC, GHG, commercial technology	Possible competition with Food
	FAME (methanol catalysis)	Biodiesel	AGR, CEff, LC, GHG, commercial technology	Possible competition with Food

High agricultural productivity: AGR; High Conversion efficiency: CEff; Low Cost: LC; Low GHG Emission: GHG

Another alternative under consideration is "green hydrogen" (H2V) which is gaining importance in Brazil. Just recently, three new production hubs were opened near the harbours of Pecém, Suape, and Açú. These three new facilities required at total investment of around US$ 20 billion.[30]

To implement a successful strategy to reduce GHG emissions in the maritime transportation sector, the basic bottom lines are the same as for the aviation sector: subsidies, compensation, or non-drop-in.

Subsidies are even less likely to succeed because no one, not even governments, will feel like to cover these costs. Besides, at least in Brazil, passengers are not the customers who will benefit most, so marketing will be much less effective.

Compensation seems to be a plausible way to go because of very large deforestation areas could benefit from a global policy to reclaim endangered areas worldwide.

Non-drop-in strategy seems also to be a good option. Probably ASTM criteria will be much less rigid for water transportation than with the aviation sector, so the development of a low-cost sustainable biofuel for the maritime sector seems to be a reachable target. Of course, the engine manufacturers will have to agree with the premises. It is important to note that the quality of fuels used in maritime transport are far less rigorous than aviation and this will be a great advantage.

In case the non-drop-in strategy becomes a plausible option, fast pyrolysis bio-oil (FPBO) obtained from low-cost fibrous biomass, could be a candidate, at least for the transition phase (CORTEZ et al, 2021). Another important route could be the use of ammonia as fuel for maritime transport, as indicated in the recent development made by Monash University from Australia.[31]

5.5.3. RAILWAY

Train is not an important sector for passenger transport in Brazil at national level. However, for load/goods transportation is relatively important, particularly for minerals such as iron ore and bauxite, oil derived products, and grains. Although the railway sector is limited in Brazil, there is a trend to grow, particularly in relatively small but important segments.[32]

In the railway sector in Brazil diesel is the most utilized fuel. The most important perspectives for biofuels in the railway sector are also related to VALE and then

30 https://www.capitalreset.com/o-brasil-na-corrida-pelo-hidrogenio-verde-com-us-20-bi-em-projetos/

31 https://www.ammoniaenergy.org/articles/three-new-australian-ammonia-production-projects/

32 https://www.gov.br/antt/pt-br/assuntos/ultimas-noticias/transporte-de-cargas-pelas-ferrovias-do-pais-cresce-30-em-marco#:~:text=Transporte%20de%20cargas%20pelas%20ferrovias%20do%20pa%C3%ADs%20cresce%2030%25%20em%20mar%C3%A7o,-Revolu%C3%A7%C3%A3o%20ferrovi%C3%A1ria%20em&text=A%20produ%C3%A7%C3%A3o%20de%20transporte%20ferrovi%C3%A1rio,mesmo%20per%C3%ADodo%20do%20ano%20passado.

PETROBRAS. Both companies have been conducting independent studies investigating the most sustainable fuels for their operations. It seems, by the nature of their business, that VALE is more likely to adopt biodiesel as part of their energy transition.

If Brazil ever implements modern railways for passenger transport, probably the model will be to use electric trains or hydrogen. Then, the main problem would be to ensure electricity comes from renewable sources adding more demand to the electricity matrix. For this reason, railways are not considered a real alternative for biofuels.

5.5.4. BIOCHEMISTRY

As it was shown on chapters 2 and 4, Brazil did encourage a biomass-based chemical industry. Table 5.11 shows the result of a survey of chemical products that could be produced from ethanol (SCHUCHARDT, 2001).

Table 5.11 Basic processes of the alcohol-chemical industry. Source: BNDES (2008) adapted from Schuchaedt (2001).

Processes	Main Products	Typical application
Dehydration	Ethene Propene Ethylene-glycol	Plastic Resins Solvents Ethyl Ether Textile Fibres
Dehydrogenation Oxygenation	Acetaldehyde	Acetic Acid Acetates Dyes
Estherification	Acetates Acrylates	Solvents Textile Fibres Adhesives
Halogenation	Ethyl chloride	Cooling Fluids Medicine Products Plastic Resins
Ammonolysis	Diethylamin Monoethylamine	Insecticide Herbicide
Dehydrogenation Dehydration	Butadiene	Synthetic Rubbers

Markets for these products are important. However, there are two factors involved: first bioethanol local price in comparison with fossil fuels. The second is the market, Brazilian and global, to accept to pay a premium, at least temporarily, until they become competitive.

Already, several enterprises have begun this pathway. Oxiteno from the Ultrapar group, uses bioethanol from sugarcane to produce ethylene since the 2080s. More recently Oxiteno has done R&DD work with UNICAMP to produce ethylene and ethylene-glycol from sugarcane cellulose pre-treated by hydrolysis.

Another Brazilian company that has explored this area is Coperbo, a Pernambuco state rubber company. Coperbo has a long history in this area, beginning as early as 1965 transforming synthetic rubber from bioethanol. Later in 1975, it created the National Alcohol-Chemical Company which was later controlled by Union Carbide and then by Dow Chemical.

Several biodegradable plastics, such as polymers that present very low carbon footprint, such it is the case of P(3HA), based on soybean oil, and P(3HB), based on glucose (AKIYAMA et al, 2003). Figure 5.5 shows a comparison with GHG emissions of plastics of fossil origin (— low density polyethylene (LDPEP), high density polyethylene (HDPE), polypropylene (PP), polystyrene (PS), and polyethylene terephthalate (b-PET)).

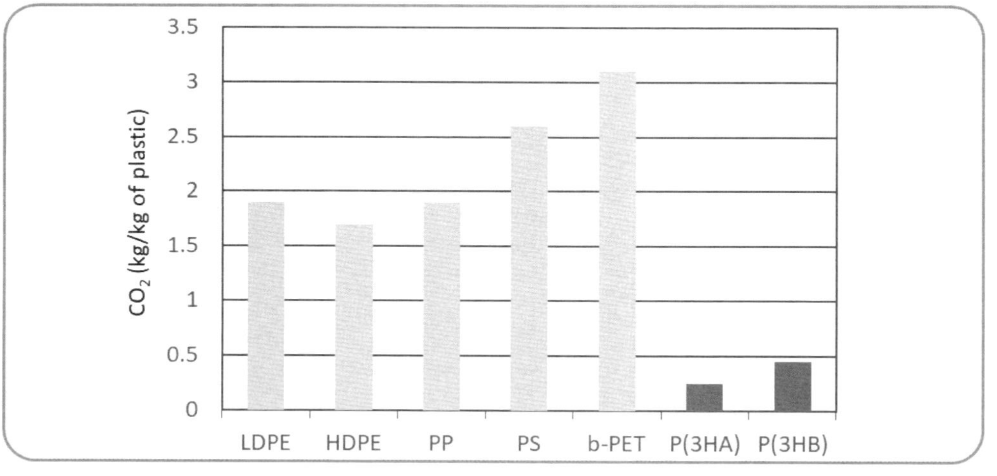

Figure 5.5 GHG emissions in the production of various plastics of fossil and biomass origin. Source: AKIYAMA et al. (2003).

More recently, Braskem has made important investments on bioplastics obtained from sugarcane bioethanol in Brazil. For further details on t biomass-based biorefineries see Macedo and Horta Nogueira (2005).

Taking into consideration current trend in the chemical sector to be as sustainable as possible, ethanol- based chemical products are bound to have a good future.

5.6. SUSTAINABILITY ISSUES AND GLOBAL COMMITMENT

Much has been said on sustainability of biofuels. Also, much of sustainability criticism is directed to feedstock and agricultural production, and land use issues.

First, it important to state that bioethanol and biodiesel production have been adapted from food crops such as sugarcane and soybeans with their own characteristics. So, it is not fair to create a series of additional requirements, e.g., on fertilizers, land use, and productivity when they were not conceived for this end. It would be more appropriate to design and implement totally new production systems, but the reality was adaptation. The development of specifically growing energy crops would be more realistic and fairer when evaluating the sustainability of biofuels.

Probably the major study carried out in Brazil on sugarcane ethanol sustainability, was coordinated by Isaias Macedo (2007). The study included 12 of the most important topics, such as: urban and rural air quality, contribution to GHG mitigation, impacts on water supply, soil occupation; new areas and biodiversity, preservation of agricultural soils, use of agricultural chemicals, use of fertilizers, sugarcane varieties, pests, and diseases control, competitivity of the Brazilian sugarcane industry, and employment and revenue.

Of course, much can be improved, but substantial gains have already been incorporated by sugarcane sector, particularly after PROALCOOL and it would be reasonable to state that Brazilian agriculture has benefitted substantially, particularly in sustainability issues and technology driven by the constant scrutiny imposed to the biofuels sector. Also, it is important to state that the Brazilian overall productivity is increasing. According EMBRAPA between 1977 and 2017, the production of grains grew from 47 to 237 million tons, nearly 5 times. Over the same period agricultural land went from 37.3 to 60.9 Mha, less than the double.[33]

FINAL REMARKS

Brazil has constructed, particularly after the oil shocks of 1970s two strong biofuels programs, one on ethanol and another for biodiesel for partial substitution with mineral fuels. Probably no other country has made proportionally such considerable effort than Brazil. Combined with a state policy to increase domestic oil production, Brazil has greatly improved its energy security while improving the environment and guaranteeing jobs internally.

However, the world scenario has changed, particularly in the last twenty years, and global warming became a great concern forcing a transition to cleaner energy. Since Brazil is a relatively small economy worldwide (circa 3% of world GDP), it is natural that the technology changes in transport will not necessarily consider the Brazilian specificities. The world is gradually adopting an energy transition that probably will conduct to different fuels n the coming two, three of four decades. The main driver to this energy transition is of course the reduction of GHG emissions, and greater energy independency.

33 https://revistacultivar.com.br/noticias/em-40-anos-agricultura-brasileira-produz-5-vezes-mais-e--area-plantada-sequer-dobra

On the other hand, biofuels produced in Brazil already generate low GHG emissions, therefore it would not be reasonable to abandon the present infrastructure and built another if there are not any environment gains. Additionally, this transformation would require significant investments in a country that needs significant investments in other important areas too. In addition, there are many direct and indirect economic and social benefits generated by biofuels which cannot be ignored without serious political consequences.

Considering these issues, together with the interests of the automobile industry, it is not clear what will be the future of biofuels in Brazil. In the best scenario, this transition will probably take few decades, giving time to the economy for a long-term adjustment and transition

As for the use of biofuels in other transportation modes, the scenario is not clear either. In all these scenarios e.g., aviation, maritime and biochemistry, except for rail transport, it seems that electrification is not the right choice. Most likely the best alternatives will favor biofuels and green fuels, such as green hydrogen.

Another alternative examined in this chapter was the possibility to integrate existing oil refining infrastructure with bioenergy feedstocks and conversion processes. Although this seems to be an attractive possibility, it is not sure the oil companies, particularly the large ones, may consider this possibility, yet.

REFERENCES

ABVE. (2021). *Eletrificados batem todas as previsões em 2021*. Available at https://www.abve.org.br/eletrificados-batem-todas-as-previsoes-em-2021/

AEA – Associação Brasileira de Engenharia Automotiva. (2017). *Eficiência Energética Sustentável*. Associação Brasileira de Engenharia Automotiva, 16 p. Available at http://antigo.mme.gov.br/documents/36224/930011/participacao_pdf_0.19998126556324192.pdf/c027c514-5c5e-1cf2-e155-acfb453b44e4.

AKIYAMA, M. et al. (2003) Environmental life cycle comparison of polyhydroxyalkanoates produced from renewable carbon resources by bacterial fermentation. *Polymer Degradation and Stability*, v. 80, issue 1, pp. 183-194. Available at https://www.sciencedirect.com/science/article/pii/S0141391002004007?via%3Dihub.

BNDES. (2008). *Sugarcane-based Bioethanol*: energy for sustainable development, BNDES-CGEE, Rio de Janeiro, Brazil, 304 p. Available at https://web.bndes.gov.br/bib/jspui/bitstream/1408/6305/1/2008_Sugarcane-based%20bioethanol_energy%20for%20sustainable_P.pdf.

BOEING/EMBRAER/FAPESP. (2013). *Flightpath to Aviation Biofuels in Brazil*: Action Plan, BOEING/EMBRAER/FAPESP Project, June 2013, 56 p.

CANALTECH. (2021). *10 carros elétricos mais vendidos no Brasil em 2021*, by AMARAL, P., jan. 7th. Available at https://canaltech.com.br/carros/carros-eletricos-mais-vendidos-brasil-2021/.

CNPEM. (2018). *Eletrificação de Veículos e o Futuro do Etanol Combustível no Brasil.* (In Portuguese), Separata CNPEM, December, 16 p.

CORTEZ, L.; FRANCO, T. T.; VALENÇA, G.; ROSILLO-CALLE, F. (2021). Perspective Use of Fast Pyrolysis Bio-Oil (FPBO) in Maritime Transport: The Case of Brazil. *Energies 2021*, 14(16), 4779. Available at https://doi.org/10.3390/en14164779

CORTEZ, L.A.B. (ed.) (2014a). *Roadmap for Sustainable Aviation Biofuels for Brazil* – a flight to aviation biofuels in Brazil, São Paulo, Blucher, 272 p. Available at http://pdf.blucher.com.br.s3-sa-east-1.amazonaws.com/openaccess/roadmap-aviation/completo.pdf.

CORTEZ, L. A. B.; SOUZA, G. M.; CRUZ, C. H. B.; MACIEL, R. (2014b). An Assessment of Brazilian Government Initiatives and Policies for the Promotion of Biofuels Through Research, Commercialization, and Private Investment Support. In *Biofuels in Brazil*, Springer, p. 31-60. Available at https://link.springer.com/book/10.1007/978-3-319-05020-1.

FOLHA de São Paulo. (2022). *Sextas Compensam*, LATAM advertisement, April 22nd.

FREIRE, G. (2011). Aviation Biofuel in Brazil. Challenges and perspectives. In: *Brazilian Bioenergy Science & Technology Conference.*

IATA – International Air Transport Association. (2021). *Annual Report*, pp. 28. Available at www.iata.org.

MACEDO, I. C.; HORTA NOGUEIRA, L. A. (2005) "Biocombustíveis". *Cadernos NAE 2*, Brasília, Centro de Gestão de Estudos Estratégicos e Núcleo de Assuntos Estratégicos (CGEE/NAE).

MELO, F. H. de; FONSECA, E. G. da. (1981). *Proálcool, Energia e Transportes*, Pioneira, 164 p.

RAÍZEN. (2022). *Sustainability of Raízen´s Ethanol*: ethanol production from sugarcane for thriving and sustainable economy. Raízen Technical Report prepared by Raízen with advisory services from Agroícone and verification by PwC (PricewaterhouseCoopers Independent Auditors), 66 p.

RODRIGUE, J. P. (2020). *The Geography of Transport Systems*, 5th ed., Routledge, New York.

SCHUCHARDT, U. et al. (2001). A indústria petroquímica no próximo século: como substituir o petróleo como matéria prima? *Química Nova*, v. 24, n. 2, pp. 247-251.

SODRE, E. (2022). Gasolina cara reduz vantagem de carros mais econômicos. (In Portuguese). Article published at *Folha de São Paulo*, April 22nd.

SZKLO, A. et al. (2020). *Perspectivas de Produção de Combustíveis Marítimos de Emissão Neutra de Carbono no Brasil.* (in Portuguese), Elaborated by Cenergia, PPE/COPPE, UFRJ, ICS, Rio de Janeiro, Brazil, 41 p.

WEI, H. K. (2021). *Análise Comparativa do Porto ao Navio de Combustíveis Alternativos para Transporte Marítimo.* MSc dissertation, COPPE, UFRJ, Rio de Janeiro, Brazil, 128 p.

CHAPTER 6
GENERAL CONCLUSIONS

As stated previously, the oil and energy sector, and more specifically transport, has not been synonymous of rapid change in the past. In fact, for decades, hardly any meaningful change took place as the automobile industry and oil giants, shared a married of convenience. However, the undergoing changes are unparalleled. There are many "unknowns" but undoubtedly such changes represent a paradigm shift with cataclysmic proportions.

Even before the Russian war on Ukraine, there was considerable uncertainly. This war has created an unpredictable geopolitical upheaval in the energy sector, exceedingly difficult to predict. One thing is sure, it will have huge economic, social, environmental and policy implications, with hundreds of billions of dollars at stage. Such changes represent both huge opportunities and also serious challenges and pitfalls.

In addition, there is even much greater challenges: climate change. The implications are enormous and beyond the scope of this book. The amount of data generated daily is such that it would be impossible to provide a reasonable representative account. One thing is sure, climate change is not something that may happen in the future, as many people might perceive it, it is already here and would not go away. It will get progressively worse if no fundamental actions are taken.

There have been many scenario predictions but hardly any of them turned out to be correct. Given the current situation, such predictions are even more uncertain. The energy sector is changing rapidly. Take, for example, the automobile of the future, it will be quite different in which decarbonization, environment and sustainable fuels will play a key role. Such changes represent an enormous challenge to both the automobile and oil industries e.g., technological, environmental, economic, political, and social.

While oil is not going away any time soon it will, progressively, be phased out over time. The Russia-Ukraine war might have two immediate impacts: i) the strengthening of fossil fuels in the short-term (2-3 years) because there is not any real alternative to fossil fuels; ii) speed up the implementation of renewable energy sources.

This is particularly the case of Western Europe whose external energy dependency is close to 90%, mostly fossil fuels. Europe consumes huge amount of energy and spend considerable financial resources. And for the first time, there is a growing political will to confront this uncomfortable reality. It is also a region with a huge human, technological, financial, and capital resources to make the transition to a sustainable energy scenario more realistic, if there is a political will to do so.

It can be stated, with certain degree of certainty, that many candidates are progressively emerging e.g., electricity, hydrogen, solar, biofuels (bioethanol, biodiesel). Therefore, the transport energy future will be dominated by a mixture of fuels in with no one playing the dominant role of oil. However, there is a danger that some countries may become too dependent on the electric car, which can lead major implications as stated.

The book main concern has been with the future of biofuels, with Brazil as the prime focus. Thus, within this context, what could be the potential role of biofuels in transport, and more specifically in Brazil? Biofuels have been around for many decades, and in particular in Brazil and USA, and to a lesser extent in other countries, hence they represent a lesser technological challenge by comparison, as it is a well proven technology and biofuel.

But as discussed in Chapter 1, the potential development of biofuels will be unequal within a global context. The Americas will continue to dominate, with Asian markets increasing their share, and with the African Continent (meaning Sub-Saharan countries), offering little future prospects, according to Maia (2022, Box 1.1). It is interesting to note that a region with considerable potential to expand biofuels, this seems unlikely in the near future due to a combination of factors e.g., social, cultural, economic.

Contrary, Görgens (see Box 1.2), is pinning up hope in advanced biofuels in conjunction with other RE (solar, hydrogen, electrification of the transport system, etc). This shows how unequally are biofuels development being perceived in different parts of the world.

***

Chapter 1 is a succinct global overview of current and future potential of biofuels in transport, subject to the uncertainties stated above. It provides a good amount of data and references for the interested reader to follow up. A point of caution regarding data. As already explained, the amount of data accumulated in so large, and changing so quickly, that often means little in the current form. Take for example, data of electric vehicle (EV), or battery technology which vary from one minute to the next. What is important is the general trend. Never in history have there been so many bright scientists and engineers, and dedicated resources, trying to find out new energy alternatives to fossil fuels.

This chapter also include a discussion of the pros and cons of biofuels and their historical perceptions. The pros and cons of biofuels have been heatedly debated, but on balance and if produced and used sustainably, the benefits far aweigh the negative impacts. Therefore, biofuels represent a considerable advantage over other emerging alternatives.

But biofuels are not excepted of negative impacts that limit their potential in many parts of the world, real or perceived. The main one being land competition with food production and the requirement of being environmentally sustainable. The "food versus fuels" dilemma, as stated, has been the object of heated debate for decades. The debate (e.g., see ROSILLO-CALLE, 2018) has been clouded by vested interest, bias, and lack of reliable scientific data. The "food versus fuel" debate needs to consider the intertwined nature of biofuels, land use, food production, food waste, animal feed production, diets, social habits, food distribution systems, policy, and so forth.

In fact, the FvF problem is an outdated, misleading, and increasingly discredited argument. Even in the EU where this argument has been a major political issue, now accept the positive contribution of biofuels to the environment and food security.[1]

The biofuels industry needs to transform itself by moving away from using feedstocks that compete directly with food production e.g., woody biomass and agroforestry residues, and explain to the wider public more successfully, as it is has not been the case so far, and with robust data, that FvF as widely portrayed so far, lack fundamental scientific credibility.

The way FvF is perceived needs then to be overcome and buried. There is even a greater problem, often overlooked, and it is the unsustainability of food production. Chapter 1 also points out that no country is an island and hence what happens around the world will impact directly or indirectly in an individual country and its policies, and Brazil is not an exception, despite its unique position and historical experience, and natural advantages with biofuels.

Environment and climate change will be key drivers in any future anergy scenario as already emphasized. This requires international agreement or some kind of consensus. But environmentalism is clouded with conflicting and contradictory vested interests. But as stated in the Introduction, when Homo Sapiens faces very serious problems, can find solutions above and over strong disagreements.

Thus, the future development of biofuels will be partly shaped by external factors. And this poses the question: *will Brazil use these advantages to promote biofuels in the future? Have biofuels a future in Brazil?* Answering these questions have been the aim of this book

***

1 See for example, David Carpintero short article: Food and Fuel for Europe – The Future of Renewable Ethanol, Ethanol Producer Magazine, 17 December 2022.

Chapter 2 deals with the historical developments of the ethanol fuel in Brazil. The history of ethanol fuel is fascinating, with many twists and turns. Ethanol fuel had had many supporters and detractors. Its origins go back to 1887 when Nikolas Otto used ethanol in his first engine. And in the USA, Gerald Ford also used ethanol in his T-Model car. In fact, Ford thought that in the future, "*cars will be powered by ethanol*". In 1907 the USADA published a very interesting paper on *The Use of Alcohol and Gasoline in Farm Engines*. It was only when oil was discovered in large quantities in the USA that gasoline became the real alternative to ethanol.

Returning to Brazil, the reader interested in the history of fuel ethanol will know that it goes back as early as 1902. During 1920s many experiments were carried out, and as can appreciated in the photo bellow (Figure 6.1), the use ethanol gasoline blend was a normal feature for automobile enthusiasts.

Figure 6.1 One of the first ethanol fuelled vehicles in Brazil, a Ford automobile, specially prepared at the Estação Experimental de Combustíveis e Minérios (today Instituto Nacional de Tecnologia-INT). Source: INT (2022).

This chapter highlights how biofuels have evolved in Brazil through its various historical phases. It has identified the most salient features such as policy changes, technological achievements, regulations, major actors in the development of biofuels, agriculture, social, and environmental issues, as well as lessons learned (e.g., see also CORTEZ, 2016).[2]

It has been a long and arduous road, with many ups and downs. It is a rich history of failures and successes. A salient feature, perhaps from the eyes of many people in the 21st century, is lack of any environmental sustainability concern. But there was a school of thought at that time, whose core policy-making philosophy was economic development. This thinking considered that social development will tricker down as a

2 The interested reader on the development of biofuels in Brazil, should read PROALCOOL 40, 1975-2015. It is in Portuguese and is an excellent account of the main protagonists on the program.

result of improving living standards, and biofuels development reflects this reality. It has been a long road since environmental sustainability has been taken seriously.

Chapter 2 provides, therefore, a realistic overview of this historical process for a spectrum of readers.

***

Chapter 3 embraces all major aspects of the current situation of biofuels in Brazil. It is a detailed description of ethanol and biodiesel development, ranging from current status, opportunities and challenges facing this industry and agricultural implications. This is back up by robust data. It is worth restating that Brazil is currently the world largest user of RE, with 48.8% of all energy, primarily hydro and biofuels.

The chapter also includes an analysis on a more recent phenomenon, that is, the use of corn to produce ethanol. This is a significant development since sugarcane has been, historically, at the core of ethanol fuel production. The chapter discusses the pros and cons of this new feedstock that can have major implications in the future. Corn is a major crop in Brazil and its use as fuel will require new models and thinking on how to integrate sugarcane-corn production. It also has major implication for agriculture because corn is intercropped with soybean. It is a new emerging reality with significant economic, technological, and social implications. Corn-fuel is expanding rapidly because of its many advantages as explained in the chapter. It is, fundamentally, a replication based on the USA experience.

The expansion of biodiesel is not without its problems. It is a more recent story compared to ethanol fuel. Biodiesel is expanding quickly, and raising some concerns because, and although many other vegetables oil can be used, currently it uses primarily soyabean (c.68%) which has expanded very quicky causing considerable environmental concern as it is linked to deforestation of the Amazon. The deforestation of tropical forests affects both climate change and global warming, and Brazil is considered the most important emitter of GHG emissions from this source. However, reality is much more complex, and the main culprit has been extensive cattle-ranching.

The chapter looks also at other emerging potential markets for biofuels- aviation, maritime transport, and chemical uses.

Land use and biofuels are often seen as incompatible. However, in Brazil given its land mass area (850Mha), with huge amount of unused or underutilized land, the FvF has never been a serious issue, as explained in the chapter.

There are many challenges ahead. For example, the sugarcane ethanol sector currently faces a considerable challenge because of a combination of factors including:

- Expanding production. In the present model sugarcane ethanol is only viable if its economics are compensated and supplemented by sugar sales
- Electric mobility. This option has been presented as a "clean" solution, although the main issue is how electricity is produced. It can only be a partial answer.

- Domestic difficulties associated with relatively low engine efficiencies. Flex-fuel engines are no longer adequately responding the consumer needs.

As biofuels in Brazil do contribute significantly to mitigate GHG emissions, it is imperative to discuss new production models to satisfy the present needs, and to connect these modern biofuels with modern/modified engines. These solutions require mid and long-term policies involving not only Brazil but also international private enterprises and governments.

Chapter 3 is a good compendium of the present situation of biofuels in Brazil. It provides a considerable amount of data.

***

As discussed, throughout this book the energy sector is in turmoil. This situation, however, will differ considerably from country to country, depending on the level of energy dependency. Fortunately for Brazil it is one the countries with considerable energy self-sufficiency and in which RE plays already a major role. Its historical experience on RE, chiefly hydro and biofuels, puts the country on a solid footing.

But having a historical experience is not enough if this is not back up by scientific, human, and technological capability, and capital. The book has also stressed the fundamental R&DD required to overcome the challenges, and Chapter 4 focuses precisely on these issues by investigating alternative biofuel models and feedstocks. To survive, requires constant change, innovation, and adaptation. We live in a world driven by this continuous change, particularly scientific and technological, driving everything around us. It is an increasing challenge compounded by its complexity.

This chapter demonstrates that Brazil has considerable R&DD capacity and know-how to develop new feedstocks, and process technology, particularly from woody biomass. Significant R&DD has been underway for many years in universities, research institutions and industry. Many routes are being investigated and it is an invigorating research field. With the right policies and support, this area could advance rapidly. Although it is also clear that there is considerable room for improvement, to move away from current unsustainable economic, technological, and environmental models.

For example, a better management of cattle ranches can easily result in freeing considerable amount of land while being environmentally more friendly. The combined production of sugarcane-corn presents an excellent social, economic, and environmental benefits while making better use of existing natural resources. This means significant changes in land use models.

There are, therefore, realistic alternatives to current unsustainable production models, ranging from process technology to new sustainable feedstocks, to economic models that make better use of existing resources with considerable economic, social, and environmental benefits.

***

As stated in previous chapters, forecasting an energy future scenario is notoriously difficult and uncertain. We live in a world driving by complex and rapid technological, social, economic, and political forces. Technological innovations and betterment, influence consumer's preferences which in turn influence scientific and technological change. A world where mas data/information is available at your fingertips, is wonderful. But not without its dark side since data can easily be manipulated to serve particular interests.

Chapter 5 tries to foresee the future of biofuels in Brazil, within this constantly changing reality, from technological adaptations of engines and fuels to possible adoption of new fuels such electric and hydrogen driven vehicles. In particular it tries to answer the following questions, specifically with regard to Brazil:

- The potential new role of biofuels in the transportation sector in a rapidly changing world energy scenario
- Assessed new emerging energy alternatives and their implications for biofuels,
- Examines potential conflicts and complementarity with biofuels of such new alternatives,
- New emerging market opportunities and challenges for biofuels.

Brazil is a unique case because of its well-developed biofuel programs (bioethanol and biodiesel) which has transcribed in considerable benefits e.g., technological, economic, and social. It is important to remember that many of the benefits from biofuel programs can be considered as "intangible". Brazilian economy and particular agriculture would not be the same without these programs.

Brazil could improve its already mixed energy matrix, with biofuels increasing their share while at the same time phasing out or reducing the contribution of fossil fuels. For example, enhancing the use of biofuels by introducing hybrid engine = ethanol + electricity, ethanol + hydrogen. The country has already in place most of the infrastructure and know-how. A few conclusions can be draw from the Brazilian experience on sustainable biofuels:

- Biofuels have become an important instrument to energy security, GHG emissions reduction, and socio-economic activity in Brazil. Greater emphasis should be given in the 21st century to decrease fossil fuels dependence while enhancing the GHG emissions agenda.
- Biofuels can play an increasingly important role in the energy transition in Brazil since they provide the greatest potential for GHG emissions reduction than any other alternative
- The FvF problem so far failed to have any significant impact in Brazil; there is not a direct relationship between biofuels and food production. On the contrary, large scale biofuels production has helped to modernize the Brazilian

agribusiness sector. In any case, it is a chimera to think biofuel production would be given priority over food production. The FvF argument is misleading and outdated.

- Land use in Brazil needs to be improved. The major issue refers to pastureland, responsible for about 20%+ of all land in the country. If pastureland and cattle production is minimally improved, substantial amount of land could revert to food and biofuels production while improving biodiversity. Land use is a key issue to protect virgin forest such as the Amazon Forest and Pantanal.
- Efforts should be made by the private sector and Federal Government to make further progress on final use of biofuels. The present model of flex-fuel engines no longer satisfies present and future needs. Also, the proposed programs RENOVABIO and ROTA 2030 should be complemented with a new automotive model.
- Biofuels can also play an important role in the Brazilian international agenda involving trade, business, and cooperation particularly with medium and low-income countries.
- Brazil has the conditions to further promote biofuels, from an economic, social, technological, and environmental perspective. The question is, is there a political will?

FINAL THOUGHT

In the objectives we asked the following: *what could be the potential role of biofuels for transport in this emerging and mixed energy scenario, and more specifically in Brazil? What can possible be the new role of biofuels in the emerging energy paradigm? What lessons can be applied from its unique Brazilian historical experience? What lessons are there for other countries?*

This book has tried to answer such questions by carrying out a detailed analysis of biofuels in Brazil, ranging from historical developments, present and potential future directions.

We are aware that some sentences/paragraphs are repeated. This is to reinforce/emphasize certain specific aspects of the subject in question.

Based on the evidence so far, it is concluded that Brazil has all the necessary conditions (human, scientific, technological, economic, social) for biofuels to thrive and enhance their future, though partly conditioned by political considerations and prevailing international conditions.

The world has moved a long way since the first biofuels programs were setup. The world that gave birth to such initiatives has long go. At the present moment we are facing very serious challenges with surpass recent decades. To start with, and unlike decades ago, environmental sustainability and clime change cannot longer be ignored. On the contrary, these are key driving forces that need to be incorporated into the new reality of biofuels. We also need to overcome the FvF argument and replace it for the multi-products benefit.

The energy sector is going through a major paradigm shift whose outcome is difficult to predict. As we have seen throughout the book, a fundamental shift is ongoing in the way we produce and use energy, away from fossil fuels and in favour of RE. We are facing a world in which no transport fuel will dominate, as did oil. Again, the current energy challenges could move in both directions: i) increasing fossil fuels (at least in the few years), and ii) accelerate RE. This energy crisis is teaching us very important lessons, one of which is that we cannot go back to old habits. It could the catapult for a truly fundamental change.

The speed in which scientific and technological innovations are being introduced is overwhelming. This is opening up many new possibilities, though not without pitfalls. The combination of many of these factors are facilitating fundamental changes on consumers attitudes.

The future of biofuels in Brazil is bright but conditioned to changes demanded by society such environmental sustainability, climate change, adopting new scientific and technological innovations, and increasing energy efficiency. It is important the political elite understands such changes and potential benefits this will bring and introduce the corresponding policies.

We are witnessing a golden era of scientific and technological innovations, which are transforming our knowledge and understanding of the world around us. In fact, so much has been accomplished and so many accumulated advances have been made, be small, that they are breath-taking. Such changes are, in many ways, beyond our capacity to absorb them. And this situation is pushing to a very narrow focus, very specialised view, in detriment of the wider reality, even for the scientific community. Take communications, for example. Do you remember when you sent a letter to your girlfriend and waited for days or weeks to have a reply? In just a few decades, to contact your girlfriend, or business partner on the other side of the world, takes just a few seconds!!

Homo Sapiens are the wonder of the world, but at the same time a scorn of the universe. In just a few centuries humans have achieved incredibly unimageable things.... The space station circles our planet, allowing to better understand of our fragile Earth; the James West telescope has gone pass the confines of our known universe.... An interminable list of great achievements... Homo Sapiens is capable of sublime incredible actions.... !!

At yet, Homo Sapiens are at the same time the scorn of the earth and perhaps of our known universe. Beyond the wonderful achievements there is a dark side. Look around you, plagued by misery, inequality, brutality, hunger, violence... Russia war against Ukraine has shown, once again, the dark side and brutality of the human animal. The human animal can achieve so many wonderful things but cannot know itself, who are we, where we are going, what is our purpose...!

Homo Sapiens so grandiose and yet so miniscule!

We live in a world full of contradictions!!

Dear reader, we have come to the end of this exciting journey. It is hoped we have succeeded in providing you with a better understanding of biofuels, specifically with regard to Brazil, past, present, and future.

REFERENCES

BRIDGMAN, O. C. (1936). Utilization of Ethanol Gasoline Blends, *Ind. Eng. Chem.*, 28 (9): 11002-1112.

CORTEZ, L. A. B. (org.) (2016). *Proálcool 40 anos*: universidades e empresas: 40 anos de ciência e tecnologia para o etanol brasileiro. São Paulo, Blucher, 224 p.

INT. (2022). *INT: Um século de inovações para o Brasil* (in Portuguese). Prepared by Maurício Moutinho; revised by Denise dos Santos Pacheco and Justo D'Avila. INT, Rio de Janeiro, Brazil, 186 p. Available at https://www.gov.br/int/pt-br/int_um-seculo-de-inovacoes-para-o-brasil_compressed.pdf.

List of Figures

List of Tables

List of Abbreviations

2DS – two-degree scenario

ABIOVE – Brazilian Association of Vegetable Oil Industries/Associação Brasileira das Indústrias de Óleos Vegetais (ABIOVE)

ABVE – Brazilian Association of Electric Cars/ Associação Brasileira do Veículo Elétrico (ABVE)

AGROPALMA – Brazilian palm enterprise

AMYRIS – G2 biofuels enterprise

ANFAVEA – National Association of Automotive Vehicle Manufacturers/ Associação Nacional dos Fabricantes de Veículos Automotores (ANFAVEA)

ANP – Brazilian National Oil Agency/ Agência Nacional de Petróleo, Gás Natural e Biocombustíveis (ANP)

APAMO – Mozambique Sugar Producers Association

ASTM – American Society of Testing of Materials

AZUL – Brazilian Airline company

B5 – 5% biodiesel in the diesel

B/l – billion litres

B/gal – billion gallons

BEN – Brazilian National Energy Balance/ Balanço Energético Nacional (BEN)

BEV – battery electric vehicles

BIOEN/FAPESP – FAPESP Bioenergy Research Program/ Programa de Bioenergia da FAPESP (BIOEN/FAPESP)

BIOPALMA – AGROPALMA – Brazilian palm enterprise

BNDES – The Brazilian Development Bank/ Banco Nacional de Desenvolvimento Econômico e Social (BNDES)

CENBIO – National Reference Center on Biomass/ Centro Nacional de Referência de Biomassa (CENBIO)

CEPEA/USP – the economic research center at Esalq/USP/ Instituto de Pesquisa Científica (CEPEA/USP)

CGEE – Brazilian National Center of Management and Strategic Studies/ Centro de Gestão de Estudos Estratégicos (CGEE)

CIDE – Brazilian Federal tax on fuels, including ethanol

CIT – Norwegian Environmental Agency

CCUS – carbon capture, utilization, and storage

COALBRA – coke and wood ethanol enterprise

CODETEC/UNICAMP – Company of Technology Development/Companhia de Desenvolvimento Tecnológico (CODETEC/UNICAMP)

COFINS – Brazilian Federal security tax on fuels, including ethanol

COGEN – energy cogeneration industry association/ Associação da Indústria de Cogeração de Energia (COGEN)

CONAB – Brazilian national agriculture supply company/ Companhia Nacional de Abastecimento (CONAB)

COPERSUCAR – Brazilian sugar & ethanol trading company

COSAN – Brazilian sugar & ethanol enterprise associated with SHELL

CPFL – Brazilian utility company/ Companhia Paulista de Força e Luz (CPFL)

CTA – Aeronautic Technology Center/ Centro Tecnológico da Aeronáutica (CTA)

CTBE – *Bioethanol* Science and Technology Laboratory/ Laboratório Nacional de Ciência e Tecnologia do Bioetanol (CTBE)

CTC – COPERSUCAR Technology Center/ Centro de Tecnologia Canavieira (CTC)

DATAGRO – strategic consulting

DDG – Dry Distiller´s Grain

DEDINI – Brazilian sugar & ethanol equipment manufacturer/ DEDINI Indústrias de Base (DEDINI)

WDG – Wet Distiller´s Grain

E100 – 100% ethanol vehicles

EACEA – Brazilian horticulture enterprise

EMBRAER – Brazilian aeronautic company

EMBRAPA – Brazilian Agricultural Research Corporation/ Empresa Brasileira Agropecuária (EMBRAPA)

ENSYN – a Canadian renewable fuels and chemical enterprise

EPA – Environment Protection Agency

EPE – Brazilian Energy Planning Agency/ Empresa Brasileira de Planejamento Energético (EPE)

EQUIPAV – Brazilian infra-structure investment group

ESALQ/USP – "Luiz de Queiroz" College of Agriculture from USP (University of São Paulo)/ Escola Superior de Agricultura Luiz de Queiroz (ESALQ/USP)

FAO – Food and Agriculture Organization

FAPESP – São Paulo State Research Funding Agency/ Fundação de Amparo à Pesquisa do Estado de São Paulo (FAPESP)

FENABRAVE – Brazilian National Federation of Vehicles/ Federação Nacional da Distribuição de Veículos Automotores (FENABRAVE)

FPBO – fast pyrolysis bio-oil

FT – Fischer-Tropsch

FvF – food versus fuel

G1, G2, G3 – first, second, and third generation biofuels

GHG – greenhouse gases

GOL – Brazilian Airline company

GranBio – Brazilian G2 ethanol producer

HDR – DEDINI Rapid Hydrolysis Process

HFO – heavy fuel oil

HEV – hybrid electric vehicles

HVO – hydrotreated vegetable oil

IAA – Brazilian Sugar and Alcohol Institute/ Instituto do Açúcar e do Álcool (IAA)

IAC – Agronomic Institute of Campinas/ Instituto Agrônomico de Campinas (IAC)

IATA – International Air Transport Association

IBAMA – Brazilian Institute for Environment and Renewable Natural Resources/ Instituto Brasileiro do Meio Ambiente e dos Recursos Naturais (IBAMA)

IBGE – Brazilian statistics enterprise/ Instituto Brasileiro de Geografia e Estatística (IBGE)

ICAO – UN International Civil Aviation Organization

ICMS – Brazilian state tax on merchandises and services/ Imposto sobre Circulação de Mercadorias e Serviços (ICMS)

IEA – International Energy Agency

iLUC – indirect land use change

IMO – International Maritime Organization

INT – Brazilian National Institute of Technology/ Instituto Nacional de Tecnologia (INT)

IPAM – Amazon environment research institute/ Instituto de Pesquisa Ambiental da Amazônia (IPAM)

IPT – Institute of Technological Research/ Instituto de Pesquisas Tecnológicas (IPT)

IRENA – International Renewable Energy Agency

ITA – Institute of Aeronautic Technology/ Instituto Tecnológico da Aeronáutica (ITA)

KAV – aviation kerosene

LATAM – Brazilian airline company

LCA – life cycle analysis

LFO – light fuel oil

LNBR – Brazilian Biorenewables National Laboratory/ Laboratório Nacional de Biorrenováveis (LNBR)

LOGUM – logistics system for ethanol transportation

LUC – land use change

MAPA – Brazilian Ministry of Agriculture/ Ministério da Agricultura, Pecuária e Abastecimento (MAPA)

MGO – marine gasoil

Mha – million hectares

MME – Brazilian Ministry of Mines and Energy/ Ministério de Minas e Energia (MME)

Mt – million tons

NOVACANA – sugar & ethanol magazine

NREL – National Research Energy Laboratory

PETROBRAS – Brazilian Oil Company

PHEV – plug-in hybrid electric vehicles

PIS/PASEP – Brazilian Federal social tax on fuels, including ethanol

PLANALSUCAR – National Program for Sugarcane Genetic Improvement/Plano Nacional de Melhoramento da Cana-de-Açúcar (PLANALSUCAR)

POLI – São Paulo University Polytechnique School/ Escola Politécnica da Universidade de São Paulo (POLI/USP)

PROALCOOL – Brazilian National Alcohol Program/ Programa Nacional do Álcool (PROALCOOL)

PROBIODIESEL – Brazilian National Biodiesel Program/ Programa Nacional do Biodiesel (PROBIODIESEL)

PROCONVE – Program for Vehicle Emission Control/ Programa de Controle da Poluição do Ar por Veículos Automotores (PROCONVE)

PV – photovoltaics

RAÍZEN – Brazilian G2 ethanol produce

R&DD- Research, development, and demonstration

RE – renewable energy

RENOVABIO – an initiative from MME to promote the production of sustainable biofuels in Brazil/

REPAR – President Getúlio Vargas Refinery/ Refinaria Presidente Getúlio Vargas (REPAR)

RFA – Renewable Fuel Association

RIDESA – interuniversitary network for the development of sugar&ethanol sector/ Rede Interuniversitária para o Desenvolvimento do Setor Sucroalcooleiro (RIDESA)

ROTA2030 – Brazilian Federal program to promote automotive engine efficiency

SAF – sustainable aviation fuels

SEEG – Brazilian GHG emissions estimative system/ Sistema de Estimativa de Emissão de Gases (SEEG)

SOFC – Solid Oxide Fuel Cell

SOLAZYME – G2 biofuels enterprise

SOV – Straight Vegetal Oil

SUZANO – a Brazilian pulp and paper enterprise

SUW – Solid Urban Wastes

tc – tons of cane

TAG – Long-Term Aspirational Goal

TERMOQUIP – a Brazilian biomass thermal conversion enterprise

TFC – total final consumption

TRS – total recoverable sugars

UFOP

UFRJ – Federal University of Rio de Janeiro/ Universidade Federal do Rio de Janeiro (UFRJ)

UNCTAD – United Nations Conference on Trade and Development

UNEM – Brazilian national corn ethanol union/ União Nacional do etanol de Milho (UNEM)

UNESP – São Paulo State University/ Universidade Estadual Paulista (UNESP)

UNICA – Brazilian Sugarcane Producers Association/ União da Indústria da Cana-de-Açúcar (UNICA)

UNICAMP – State University of Campinas/ Universidade Estadual de Campinas (UNICAMP)

USDA – United States Department of Agriculture
USP – University of São Paulo/ Universidade de São Paulo (USP)
VALE – Brazilian multinational operating in the mining sector

About the authors

LUÍS AUGUSTO BARBOSA CORTEZ

Former Professor from the School of Agricultural Engineering at the State University of Campinas (UNICAMP) from 1988 to 2017. Presently Senior Researcher associated to the Energy Planning Center-NIPE-UNICAMP.

Coordinated the Energy Planning Center-NIPE-UNICAMP between 1997-2002, and 2012-2013. Coordinator of International Relations of UNICAMP between 2002-2009 and Vice-Rector of International Relations of UNICAMP 2013-2017. Co-coordinated projects in Bioenergy such as a study on sugarcane ethanol expansion in Brazil to substitute 10% of the world gasoline consumption. Administration Coordinator at ILUM School of Sciences at the Brazilian Center for Research in Energy and Materials-CNPEM, Brazil (2021-2023). Coordinated books in bioenergy such as: "Sugarcane Bioethanol: R&D on Sustainability and Productivity" published in 2010 (Jabuti Prize); "Roadmap on Sustainable Biofuels for Aviation in Brazil" – Boeing/Embraer Project, book published in 2014; "Sugarcane Bioenergy for Sustainable Development in Latin America and Africa" published in 2019.

FRANK ROSILLO-CALLE[1]

Frank Rosillo-Calle is an Honorary Senior Research Fellow, Imperial College London, UK.

He has over 40 years research experience on biomass energy, with particular interest in Brazil. He has been involved in many international research projects and has

1 In publications, the author's name appears either as Frank or Francisco.

published extensively in scientific journals and books e.g., Contributor/Editor of Energy from Alcohol (1983), The Charcoal Dilemma (1996), Industrial Uses of Energy from Biomass (2000), Biomass Energy Handbook (2007 and 2015). He has taught biomass/biofuel energy courses at various universities in Brazil, Spain, UK, and Thailand.